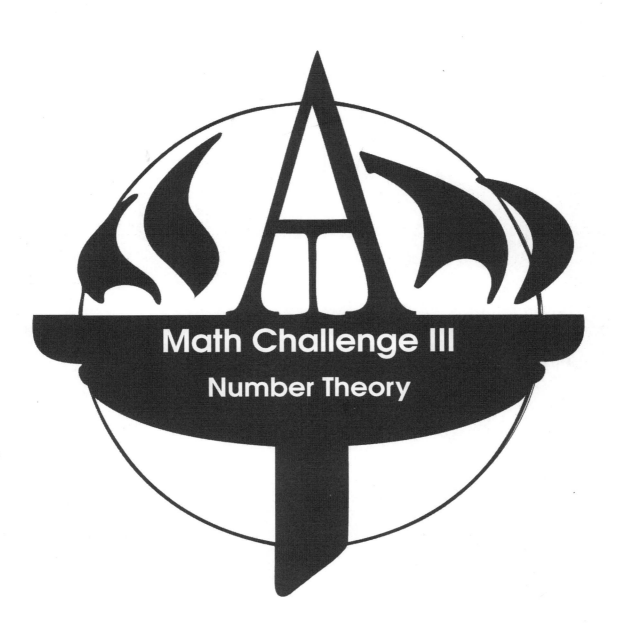

Math Challenge III
Number Theory

Areteem Institute

Math Challenge III Number Theory

Edited by   Kevin Wang
David Reynoso
John Lensmire
Kelly Ren

ISBN: 1-944863-43-5
ISBN-13: 978-1-944863-43-2
First printing, February 2019.

TITLES PUBLISHED BY ARETEEM PRESS

Cracking the High School Math Competitions (and Solutions Manual) - Covering AMC 10 & 12, ARML, and ZIML
Mathematical Wisdom in Everyday Life (and Solutions Manual) - From Common Core to Math Competitions
Geometry Problem Solving for Middle School (and Solutions Manual) - From Common Core to Math Competitions
Fun Math Problem Solving For Elementary School (and Solutions Manual)

ZIML MATH COMPETITION BOOK SERIES

ZIML Math Competition Book Division E 2016-2017
ZIML Math Competition Book Division M 2016-2017
ZIML Math Competition Book Division H 2016-2017
ZIML Math Competition Book Jr Varsity 2016-2017
ZIML Math Competition Book Varsity Division 2016-2017
ZIML Math Competition Book Division E 2017-2018
ZIML Math Competition Book Division M 2017-2018
ZIML Math Competition Book Division H 2017-2018
ZIML Math Competition Book Jr Varsity 2017-2018
ZIML Math Competition Book Varsity Division 2017-2018

MATH CHALLENGE CURRICULUM TEXTBOOKS SERIES

Math Challenge I-A Pre-Algebra and Word Problems
Math Challenge I-B Pre-Algebra and Word Problems
Math Challenge I-C Algebra
Math Challenge II-A Algebra
Math Challenge II-B Algebra
Math Challenge III Algebra
Math Challenge I-A Geometry
Math Challenge I-B Geometry
Math Challenge I-C Topics in Algebra
Math Challenge II-A Geometry
Math Challenge II-B Geometry
Math Challenge III Geometry
Math Challenge I-A Counting and Probability
Math Challenge I-B Counting and Probability
Math Challenge I-C Geometry

Math Challenge II-A Combinatorics
Math Challenge II-B Combinatorics
Math Challenge III Combinatorics
Math Challenge I-A Number Theory
Math Challenge I-B Number Theory
Math Challenge I-C Finite Math
Math Challenge II-A Number Theory
Math Challenge II-B Number Theory
Math Challenge III Number Theory

COMING SOON FROM ARETEEM PRESS

Fun Math Problem Solving For Elementary School Vol. 2 (and Solutions Manual)
Counting & Probability for Middle School (and Solutions Manual) - From Common Core to Math Competitions
Number Theory Problem Solving for Middle School (and Solutions Manual) - From Common Core to Math Competitions

The books are available in paperback and eBook formats (including Kindle and other formats).
To order the books, visit https://areteem.org/bookstore.

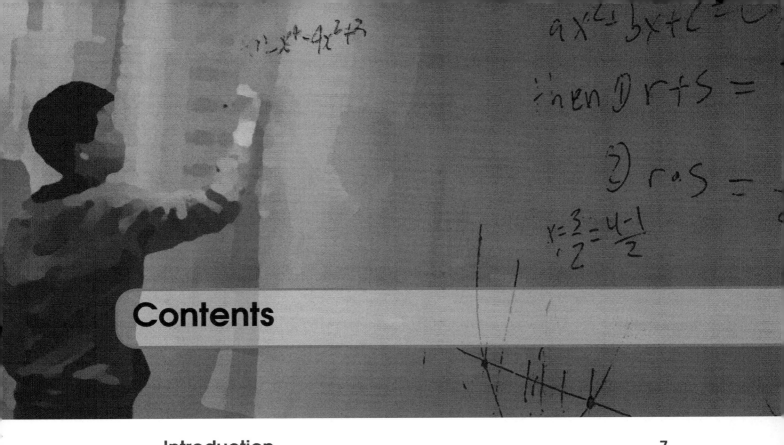

Contents

Solutions to the Example Questions 69

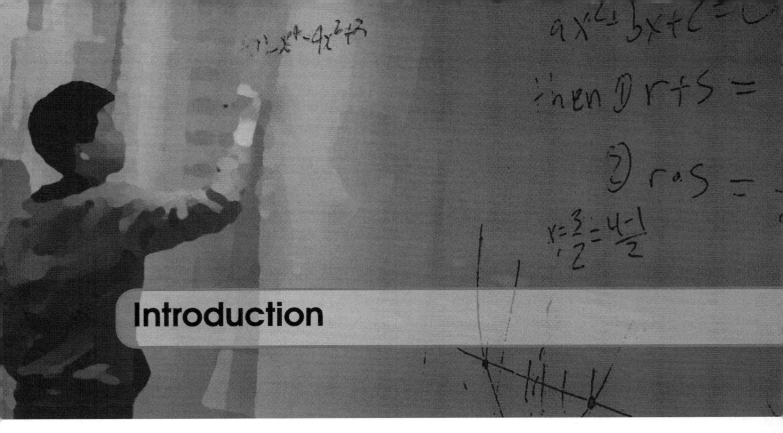

Introduction

The math challenge curriculum textbook series is designed to help students learn the fundamental mathematical concepts and practice their in-depth problem solving skills with selected exercise problems. Ideally, these textbooks are used together with Areteem Institute's corresponding courses, either taken as live classes or as self-paced classes. According to the experience levels of the students in mathematics, the following courses are offered:

- Fun Math Problem Solving for Elementary School (grades 3-5)
- Algebra Readiness (grade 5; preparing for middle school)
- Math Challenge I-A Series (grades 6-8; intro to problem solving)
- Math Challenge I-B Series (grades 6-8; intro to math contests e.g. AMC 8, ZIML Div M)
- Math Challenge I-C Series (grades 6-8; topics bridging middle and high schools)
- Math Challenge II-A Series (grades 9+ or younger students preparing for AMC 10)
- Math Challenge II-B Series (grades 9+ or younger students preparing for AMC 12)
- Math Challenge III Series (preparing for AIME, ZIML Varsity, or equivalent contests)
- Math Challenge IV Series (Math Olympiad level problem solving)

These courses are designed and developed by educational experts and industry professionals to bring real world applications into the STEM education. These programs are ideal for students who wish to win in Math Competitions (AMC, AIME, USAMO, IMO,

ARML, MathCounts, Math League, Math Olympiad, ZIML, etc.), Science Fairs (County Science Fairs, State Science Fairs, national programs like Intel Science and Engineering Fair, etc.) and Science Olympiad, or purely want to enrich their academic lives by taking more challenges and developing outstanding analytical, logical thinking and creative problem solving skills.

The Math Challenge III (MC III) courses are for students who are qualified to participate in the AIME contest, or at the equivalent level of experience. The MC III topics include polynomials, inequalities, special algebraic techniques, triangles and polygons, coordinates, numbers and divisibility, modular arithmetic, advanced counting strategies, binomial coefficients, sequence and series, complex numbers, trigonometry, logarithms, and various other topics, and the focus is more on in-depth problem solving strategies, including pairing, change of variables, advanced techniques in number theory and combinatorics, advanced probability theory and techniques, geometric transformations, etc. The curricula have been proven to help students develop strong problem solving skills that make them perform well in math contests such as AIME, ZIML, and ARML.

The course is divided into four terms:

- Summer, covering Algebra
- Fall, covering Geometry
- Winter, covering Combinatorics
- Spring, covering Number Theory

The book contains course materials for Math Challenge III: Number Theory.

We recommend that students take all four terms. Each of the individual terms is self-contained and does not depend on other terms, so they do not need to be taken in order, and students can take single terms if they want to focus on specific topics.

Students can sign up for the course at `classes.areteem.org` for the live online version or at `edurila.com` for the self-paced version.

About Areteem Institute

Areteem Institute is an educational institution that develops and provides in-depth and advanced math and science programs for K-12 (Elementary School, Middle School, and High School) students and teachers. Areteem programs are accredited supplementary programs by the Western Association of Schools and Colleges (WASC). Students may attend the Areteem Institute in one or more of the following options:

- Live and real-time face-to-face online classes with audio, video, interactive online whiteboard, and text chatting capabilities;
- Self-paced classes by watching the recordings of the live classes;
- Short video courses for trending math, science, technology, engineering, English, and social studies topics;
- Summer Intensive Camps held on prestigious university campuses and Winter Boot Camps;
- Practice with selected free daily problems and monthly ZIML competitions at ziml.areteem.org.

Areteem courses are designed and developed by educational experts and industry professionals to bring real world applications into STEM education. The programs are ideal for students who wish to build their mathematical strength in order to excel academically and eventually win in Math Competitions (AMC, AIME, USAMO, IMO, ARML, MathCounts, Math Olympiad, ZIML, and other math leagues and tournaments, etc.), Science Fairs (County Science Fairs, State Science Fairs, national programs like Intel Science and Engineering Fair, etc.) and Science Olympiads, or for students who purely want to enrich their academic lives by taking more challenging courses and developing outstanding analytical, logical, and creative problem solving skills.

Since 2004 Areteem Institute has been teaching with methodology that is highly promoted by the new Common Core State Standards: stressing the conceptual level understanding of the math concepts, problem solving techniques, and solving problems with real world applications. With the guidance from experienced and passionate professors, students are motivated to explore concepts deeper by identifying an interesting problem, researching it, analyzing it, and using a critical thinking approach to come up with multiple solutions.

Thousands of math students who have been trained at Areteem have achieved top honors and earned top awards in major national and international math competitions, including Gold Medalists in the International Math Olympiad (IMO), top winners and qualifiers at the USA Math Olympiad (USAMO/JMO) and AIME, top winners at the

Zoom International Math League (ZIML), and top winners at the MathCounts National Competition. Many Areteem Alumni have graduated from high school and gone on to enter their dream colleges such as MIT, Cal Tech, Harvard, Stanford, Yale, Princeton, U Penn, Harvey Mudd College, UC Berkeley, or UCLA. Those who have graduated from colleges are now playing important roles in their fields of endeavor.

Further information about Areteem Institute, as well as updates and errata of this book, can be found online at http://www.areteem.org.

About Zoom International Math League

The Zoom International Math League (ZIML) has a simple goal: provide a platform for students to build and share their passion for math and other STEM fields with students from around the globe. Started in 2008 as the Southern California Mathematical Olympiad, ZIML has a rich history of past participants who have advanced to top tier colleges and prestigious math competitions, including American Math Competitions, MATHCOUNTS, and the International Math Olympaid.

The ZIML Core Online Programs, most available with a free account at ziml.areteem.org, include:

- **Daily Magic Spells:** Provides a problem a day (Monday through Friday) for students to practice, with full solutions available the next day.
- **Weekly Brain Potions:** Provides one problem per week posted in the online discussion forum at ziml.areteem.org. Usually the problem does not have a simple answer, and students can join the discussion to share their thoughts regarding the scenarios described in the problem, explore the math concepts behind the problem, give solutions, and also ask further questions.
- **Monthly Contests:** The ZIML Monthly Contests are held the first weekend of each month during the school year (October through June). Students can compete in one of 5 divisions to test their knowledge and determine their strengths and weaknesses, with winners announced after the competition.
- **Math Competition Practice:** The Practice page contains sample ZIML contests and an archive of AMC-series tests for online practice. The practices simulate the real contest environment with time-limits of the contests automatically controlled by the server.
- **Online Discussion Forum:** The Online Discussion Forum is open for any comments and questions. Other discussions, such as hard Daily Magic Spells or the Weekly Brain Potions are also posted here.

These programs encourage students to participate consistently, so they can track their progress and improvement each year.

In addition to the online programs, ZIML also hosts onsite Local Tournaments and Workshops in various locations in the United States. Each summer, there are onsite ZIML Competitions at held at Areteem Summer Programs, including the National ZIML Convention, which is a two day convention with one day of workshops and one day of competition.

ZIML Monthly Contests are organized into five divisions ranging from upper elementary school to advanced material based on high school math.

- **Varsity:** This is the top division. It covers material on the level of the last 10 questions on the AMC 12 and AIME level. This division is open to all age levels.
- **Junior Varsity:** This is the second highest competition division. It covers material at the AMC 10/12 level and State and National MathCounts level. This division is open to all age levels.
- **Division H:** This division focuses on material from a standard high school curriculum. It covers topics up to and including pre-calculus. This division will serve as excellent practice for students preparing for the math portions of the SAT or ACT. This division is open to all age levels.
- **Division M:** This division focuses on problem solving using math concepts from a standard middle school math curriculum. It covers material at the level of AMC 8 and School or Chapter MathCounts. This division is open to all students who have not started grade 9.
- **Division E:** This division focuses on advanced problem solving with mathematical concepts from upper elementary school. It covers material at a level comparable to MOEMS Division E. This division is open to all students who have not started grade 6.

The ZIML site features are also provided on the ZIML Mobile App, which is available for download from Apple's App Store and Google Play Store.

Acknowledgments

This book contains many years of collaborative work by the staff of Areteem Institute. This book could not have existed without their efforts. Huge thanks go to the Areteem staff for their contributions!

The examples and problems in this book were either created by the Areteem staff or adapted from various sources, including other books and online resources. Especially, some good problems from previous math competitions and contests such as AMC, AIME, ARML, MATHCOUNTS, and ZIML are chosen as examples to illustrate concepts or problem-solving techniques. The original resources are credited whenever possible. However, it is not practical to list all such resources. We extend our gratitude to the original authors of all these resources.

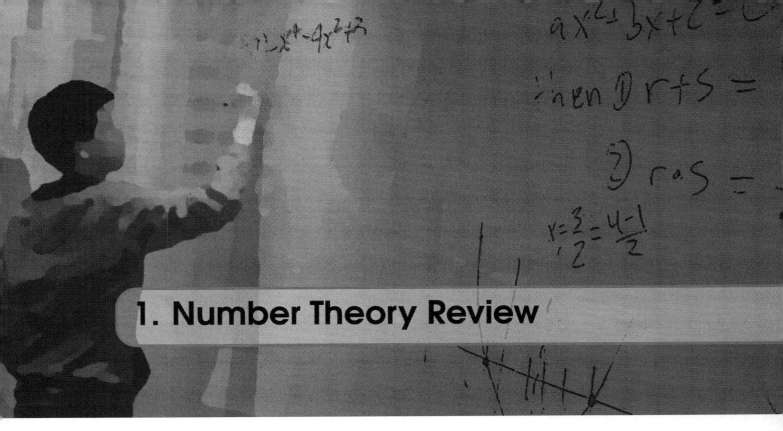

1. Number Theory Review

In this chapter we review the fundamental concepts and facts you should know in number theory.

- **Divisibility**

Divisor and Multiple

If an integer a divides an integer n evenly (that is, there is an integer c such that $ac = n$), then a is a divisor (or a factor) of n, we write $a \mid n$ to denote that a divides n. Also we say n is a multiple of a.

Prime Number

A number $p > 1$ is prime if its only divisors are itself and 1.

Theorem 1.1 Unique Factorization

Every positive integer n has a unique factorization as a product of primes: $n = p_1^{e_1} p_2^{e_2} \cdots p_k^{e_k}$ for distinct primes p_i and $e_i > 0$.

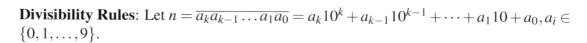

Divisibility Rules: Let $n = \overline{a_k a_{k-1} \ldots a_1 a_0} = a_k 10^k + a_{k-1} 10^{k-1} + \cdots + a_1 10 + a_0, a_i \in \{0, 1, \ldots, 9\}$.

(a) By 2 : If the last digit of n is even $(0, 2, 4, 6, 8)$, then $2 \mid n$.

(b) By 3: Let $S(n)$ represent the sum of n's digits. If $3 \mid S(n)$, then $3 \mid n$.

(c) By 4: if the last two digits of n is divisible by 4, then $4 \mid n$.

(d) By 5: If the last digit of n is 0 or 5, then $5 \mid n$.

(e) By 6: check both 2 and 3.

(f) By 8: if the last three digits of n is divisible by 8, then $8 \mid n$.

(g) By 9: Let $S(n)$ represent the sum of n's digits. If $9 \mid S(n)$, then $9 \mid n$.

(h) By 11: If $11 \mid (a_0 - a_1 + a_2 - \cdots + (-1)^k a_k)$, then $11 \mid n$.

(i) By 7 and 13: Use the fact that $1001 = 7 \times 11 \times 13$. **Example:** 15197. Split out the right-most 3 digits, so we have 15 and 197. Subtract the smaller from the larger, in this case $197 - 15 = 182$. Doing division, 182 is a multiple of both 7 and 13. Therefore 15197 is a multiple of both 7 and 13.

For numbers with more than 6 digits: split the rightmost 3 digit in each step and do multiple steps.

Number of Factors

For an integer n, $\tau(n)$ is the number of factors of n. If $n = p_1^{e_1} p_2^{e_2} \cdots p_k^{e_k}$ is the prime factorization of n, then

$$\tau(n) = \prod_{i=1}^{k} (e_i + 1).$$

Example 1.1

Consider the number $200 = 2^3 \times 5^2$. The exponents are 3 and 2, so the number of factors is $(3+1)(2+1) = 12$. Thus 200 has 12 factors. We can list them: 1, 2, 4, 5, 8, 10, 20, 25, 40, 50, 100, 200. A better way is to list them in the following table:

	2^0	2^1	2^2	2^3
5^0	1	2	4	8
5^1	5	10	20	40
5^2	25	50	100	200

Greatest Common Divisor and Least Common Multiple

Let m and n be positive integers, then the greatest common divisor (GCD) of m and n (denoted $\gcd(m,n)$) is the largest number d such that $d \mid m$ and $d \mid n$, and the least common multiple (LCM) of m and n (denoted $\mathrm{lcm}(m,n)$) is the smallest number L such that $m \mid L$ and $n \mid L$.

Properties of GCD and LCM

Suppose $m = p_1^{e_1} p_2^{e_2} \cdots p_k^{e_k}$ and $n = p_1^{d_1} p_2^{d_2} \cdots p_k^{d_k}$ (here, e_i and d_i can equal zero).

(a) $\gcd(m,n) = \prod_{i=1}^{k} p_i^{\min(d_i,e_i)}$.

(b) $\mathrm{lcm}(m,n) = \prod_{i=1}^{k} p_i^{\max(d_i,e_i)}$.

(c) Two integers m and n are relatively prime if $\gcd(m,n) = 1$.

(d) If $a \mid m$ and $a \mid n$, then $a \mid \gcd(m,n)$.

(e) If $m \mid k$ and $n \mid k$, then $\mathrm{lcm}(m,n) \mid k$.

(f) $mn = \gcd(m,n) \times \mathrm{lcm}(m,n)$. (Prove it!)

Division Algorithm

For any integers a and b, $a \neq 0$, there exists a unique pair (q, r) of integers such that $b = aq + r$ and $0 \leq r < |a|$. The number q is called the *quotient*, and r is called the *remainder*.

Euclidean Algorithm

Assume $m \geq n$ and let $r_0 = m$ and $r_1 = n$, then there exists sequences of integers $q_1, \ldots, q_k$ and $r_2, \ldots, r_k$ such that:

$$
\begin{aligned}
r_0 &= r_1 q_1 + r_2 \quad \text{with} \quad 0 < r_2 < r_1 \\
r_1 &= r_2 q_2 + r_3 \quad \text{with} \quad 0 < r_3 < r_2 \\
&\vdots \qquad\qquad\qquad\qquad \vdots \\
r_{k-1} &= r_k q_k \qquad\quad \text{with} \quad r_k = \gcd(m, n)
\end{aligned}
$$

Example 1.2

We find the greatest common divisor of 900 and 243 using the Euclidean algorithm:

$$
\begin{aligned}
900 &= 243 \times 3 + 171, \\
243 &= 171 \times 1 + 72, \\
171 &= 72 \times 2 + 27, \\
72 &= 27 \times 2 + 18, \\
27 &= 18 \times 1 + 9, \\
18 &= 9 \times 2. \qquad \text{(Remainder is 0, so the last divisor 9 is our answer)}
\end{aligned}
$$

Therefore $\gcd(900, 243) = 9$.

> ### Theorem 1.2 Bézout's Identity
>
> For any two positive integers m and n, there exist integers a and b such that $am + bn = \gcd(m, n)$.

The proof of this theorem is beyond the scope of this class. Nevertheless, you may use this result when applicable.

- **Place Values and Number Bases**

> ### Place Values
>
> The value of a digit depends on its place, or position, in a number. For example,
>
> $$796 = 7 \times 100 + 9 \times 10 + 6$$
> $$2013 = 2 \times 1000 + 1 \times 10 + 3$$
>
> In general,
>
> $$\overline{abc} = a \times 100 + b \times 10 + c$$
> $$\overline{a_1 a_2 \cdots a_n} = a_1 \times 10^{n-1} + a_2 \times 10^{n-2} + \cdots + a_{n-1} \times 10 + a_n$$

Different Number Bases

In the example above we used base 10, also called the *decimal system*. Other positive integers (other than 1) can be used as bases as well. Numbers in other bases are expressed in the same manner as the decimal system. For example, the binary system has base 2, and only uses two symbols (0 and 1) to represent numbers, and each of the binary digits also has their place values: units place, 2s place, 4s place, etc.

Usually, the base number is written as subscripts, such as 101_2. Here, for decimal numbers we omit the base number, that is, we simply write 345 instead of 345_{10}.

Conversion to Decimal from Other Bases

(a) Binary (base 2) numbers: $1101_2 = 1 \times 2^3 + 1 \times 2^2 + 0 \times 2 + 1 = 8 + 4 + 1 = 13$

(b) Octal (base 8) numbers: $567_8 = 5 \times 8^2 + 6 \times 8 + 7 = 375$

(c) Hexadecimal (base 16) numbers: use A, B, C, D, E, F to represent 10, 11, 12, 13, 14, 15:

$$
\begin{aligned}
& 57C09AF_{16} \\
=\ & 5 \times 16^6 + 7 \times 16^5 + 12 \times 16^4 + 0 \times 16^3 + 9 \times 16^2 + 10 \times 16 + 15 \\
=\ & 92015023
\end{aligned}
$$

Conversion from Decimal to Binary

(a) For the integer part: divide by 2, and record the remainder (either 0 or 1) at each step, and repeat. After reaching 0, write out the remainders in reverse order.

(b) For the fractional part: keep multiplying by 2 and taking the integer parts (also either 0 or 1) as the digits following the dot.

Example 1.3

Convert 13 to binary:

$$
\begin{aligned}
13 \div 2 &= 6 \ \cdots \ 1 \\
6 \div 2 &= 3 \ \cdots \ 0 \\
3 \div 2 &= 1 \ \cdots \ 1 \\
1 \div 2 &= 0 \ \cdots \ 1
\end{aligned}
$$

So $13_{10} = 1101_2$.

Example 1.4

Convert 0.6875 to binary:

$$
\begin{aligned}
0.6875 \times 2 &= 1.375 &\cdots\text{take} \quad 1 \\
0.375 \times 2 &= 0.75 &\cdots\text{take} \quad 0 \\
0.75 \times 2 &= 1.5 &\cdots\text{take} \quad 1 \\
0.5 \times 2 &= 1 &\cdots\text{take} \quad 1
\end{aligned}
$$

So $0.6875_{10} = 0.1011_2$.

Operations in Other Bases

Operations in other bases work the same way as in decimal system, just keep in mind that carry occurs at the base number (for example, in binary system, carry occurs whenever you have "2").

Example 1.5

In the decimal (base 10) system, $38 + 13 = 51$. Convert the summands to binary: $38 = 100110_2$, and $13 = 1101_2$. Let's perform binary addition (remember $1 + 1 = 10$ in the binary system):

$$
\begin{array}{ccccccc}
 & 1 & 0 & 0 & 1 & 1 & 0 \\
+ & & & 1 & 1 & 0 & 1 \\
\hline
 & 1 & 1 & 0 & 0 & 1 & 1
\end{array}
$$

The result 110011_2 is equivalent to 51 in base 10.

• **Modular Arithmetic**

Modular Congruence

Two numbers a and b are *congruent modulo m* (denoted $a \equiv b \pmod{m}$) if $m \mid (a-b)$.

Properties of the modular congruence relation

(a) If $a \equiv b \pmod{m}$ and $b \equiv c \pmod{m}$, then $a \equiv c \pmod{m}$.

(b) If $a \equiv b \pmod{m}$ and $c \equiv d \pmod{m}$, then $(a+c) \equiv (b+d) \pmod{m}$.

(c) If $a \equiv b \pmod{m}$ and $c \equiv d \pmod{m}$, then $ac \equiv bd \pmod{m}$.

(d) If $a \equiv b \pmod{m}$ and $d \mid m$, then $a \equiv b \pmod{d}$.

(e) If (and only if) a and m are relatively prime, then there exists an integer $b < m$ such that $ab \equiv 1 \pmod{m}$. The number b is called the *modular multiplicative inverse* of a modulo m.

Proof of (e). Consider the set of numbers $a, 2a, 3a, \ldots, (m-1)a$. We claim that they are all distinct modulo m. Suppose $k_1 a \equiv k_2 a \pmod{m}$, then $(k_1 - k_2)a \equiv 0 \pmod{m}$. Since $\gcd(a, m) = 1$, $k_1 - k_2$ is a multiple of m. Combining with the fact that $1 \leq k_1, k_2 \leq m-1$, hence $k_1 \equiv k_2 \pmod{m}$. Therefore if $k_1 \neq k_2$, $k_1 a \not\equiv k_2 a \pmod{m}$. So $a, 2a, 3a, \ldots, (m-1)a$ are all distinct modulo m, and none of them is a multiple of m, thus one of them must be $1 \pmod{m}$. Let this number be b, so $ab \equiv 1 \pmod{m}$, that is, the modular multiplicative inverse of a exists if $\gcd(a, m) = 1$. ∎

Residue Classes

Integers a and b are said to be in the same *residue class modulo m* if $a \equiv b \pmod{m}$.

Complete Residue System

A set S of integers is called a *complete set of residue classes modulo m* if for each $0 \leq i \leq m-1$, there is an element $s \in S$ such that $i \equiv s \pmod{m}$. Also called a *complete residue system modulo m*.

(a) For any integer a, $\{a, a+1, a+2, \ldots, a+m-1\}$ is a complete set of residue classes modulo m.

(b) $\{0, 1, \ldots, m-1\}$ is the *minimal nonnegative complete set of residue classes* modulo m.

(c) It is common to consider the complete set of residue classes $\{0, \pm 1, \pm 2, \ldots, \pm k\}$ for $m = 2k+1$ and $\{0, \pm 1, \pm 2, \ldots, \pm(k-1), k\}$ for $m = 2k$.

Reduced Residue System

A set S of integers is called a *reduced set of residue classes modulo m* if for each $0 \leq i \leq m-1$ where $\gcd(i, m) = 1$, there is an element $s \in S$ such that $i \equiv s \pmod{m}$. Also called a *reduced residue system modulo m*.

(a) If p is prime, $\{1, 2, \ldots, p-1\}$ is a reduced set of residue classes modulo p.

(b) The number of elements in a reduced set of residue classes modulo m is denoted as $\phi(m)$. This is known as the *Euler ϕ function*, also called the *Euler totient function*.

Example 1.6

Let $m = 10$, then $\{1, 3, 7, 9\}$ is a reduced set of residue classes modulo 10, and $\phi(10) = 4$.

Theorem 1.3 Fermat's Little Theorem

If p is prime and a is any integer, then $p \mid (a^p - a)$. Equivalently, if p does not divide a, then $a^{p-1} \equiv 1 \pmod{p}$.

Proof. We apply a combinatorial method to prove that $p \mid (a^p - a)$. Let p be a prime and a be a positive integer such that $\gcd(a, p) = 1$ (the case where a is a multiple of p is trivial). Suppose we have beads of various colors and wish to make a bracelet consisting of p beads. Assume a is the number of the colors, and there are ample supply of beads of each color. If we do not rotate the bracelet, each position has a choices and the total number of possible bracelets is a^p. If we consider the bracelets whose only difference is a rotation as identical bracelets, then each bracelet is counted p times unless all of the beads are of the same color. The number of uni-color bracelets is a, thus $a^p - a$ must be divisible by p, which is what we wanted.   ∎

A more conventional proof is given next.

Proof. Consider the reduced set of residue classes $a, 2a, 3a, \ldots, (p-1)a$ modulo p. This set covers all the non-zero remainders modulo p, thus

$$(a)(2a)(3a) \cdots ((p-1)a) \equiv (p-1)! \pmod{p}.$$

So we get $(p-1)! \cdot a^{p-1} \equiv (p-1)! \pmod{p}$. Clearly $(p-1)!$ and p are relatively prime. Multiplying the modular multiplicative inverse of $(p-1)!$, we get $a^{p-1} \equiv 1 \pmod{p}$.   ∎

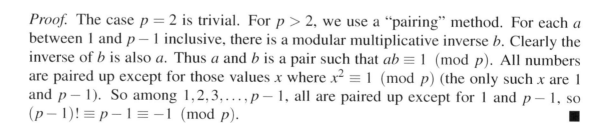

Theorem 1.4 Wilson's Theorem

For every prime p, $(p-1)! \equiv -1 \pmod{p}$.

Proof. The case $p = 2$ is trivial. For $p > 2$, we use a "pairing" method. For each a between 1 and $p-1$ inclusive, there is a modular multiplicative inverse b. Clearly the inverse of b is also a. Thus a and b is a pair such that $ab \equiv 1 \pmod{p}$. All numbers are paired up except for those values x where $x^2 \equiv 1 \pmod{p}$ (the only such x are 1 and $p-1$). So among $1, 2, 3, \ldots, p-1$, all are paired up except for 1 and $p-1$, so $(p-1)! \equiv p-1 \equiv -1 \pmod{p}$.   ∎

Theorem 1.5 Chinese Remainder Theorem

Let $m_1,\ldots,m_k$ be pairwise relatively prime positive integers (that is, $\gcd(m_i,m_j)=1$ for all $i \neq j$). Let $b_1,\ldots,b_k$ be arbitrary integers. Then the system

$$
\begin{aligned}
x &\equiv b_1 &&(\bmod\ m_1) \\
x &\equiv b_2 &&(\bmod\ m_2) \\
&\ \ \vdots \\
x &\equiv b_k &&(\bmod\ m_k)
\end{aligned}
$$

has a unique solution modulo $m_1 m_2 \cdots m_k$.

For the Chinese Remainder Theorem, we shall not give a general proof, but use an example to illustrate how the solution is found, and the method is easily converted to a proof of the general case.

Example 1.7

(Based on ancient Chinese text.) An army general was counting his soldiers. The total number of soldiers was between 100 and 200. He let the soldiers stand in rows of 3 each. There were 2 soldiers left out. He changed the row size to 5 soldiers each, then there were 3 soldiers left out. Finally he changed to 7 soldiers per row, and there were 2 soldiers left out. How many soldiers were there?

Solution

This problem is equivalent to the following system of modular equations:

$$
\begin{aligned}
x &\equiv 2 &&(\bmod\ 3) \\
x &\equiv 3 &&(\bmod\ 5) \\
x &\equiv 2 &&(\bmod\ 7)
\end{aligned}
$$

In the Chinese text, the solution was described in a poem of 4 lines. The first line of the poem says (translated into modular arithmetic language): multiply the remainder mod 3 by **70**; in this case, $2 \times 70 = 140$.

The second line says: multiply the remainder mod 5 by **21**; so $3 \times 21 = 63$.
The third line says: multiply the remainder mod 7 by **15**; so $2 \times 15 = 30$.
The fourth line says: add them all up and subtract a multiple of 105; that is $140 + 63 + 30 - 105 = 128$ (choose 128 because it was given that the number is between 100 and 200).
Thus there were 128 soldiers.

How did the solution work? Let's see what the three highlighted numbers (70, 21, and 15) are in the 3 moduli:

$$
\begin{array}{lll}
70 \equiv 1 \pmod 3 & 21 \equiv 0 \pmod 3 & 15 \equiv 0 \pmod 3 \\
70 \equiv 0 \pmod 5 & 21 \equiv 1 \pmod 5 & 15 \equiv 0 \pmod 5 \\
70 \equiv 0 \pmod 7 & 21 \equiv 0 \pmod 7 & 15 \equiv 1 \pmod 7
\end{array}
$$

If each number is multiplied by the corresponding remainder (as specified in the poem), we get

$$
\begin{array}{lll}
2 \times 70 \equiv 2 \pmod 3 & 3 \times 21 \equiv 0 \pmod 3 & 2 \times 15 \equiv 0 \pmod 3 \\
2 \times 70 \equiv 0 \pmod 5 & 3 \times 21 \equiv 3 \pmod 5 & 2 \times 15 \equiv 0 \pmod 5 \\
2 \times 70 \equiv 0 \pmod 7 & 3 \times 21 \equiv 0 \pmod 7 & 2 \times 15 \equiv 2 \pmod 7
\end{array}
$$

Therefore, when we add up the products,

$$
\begin{array}{ll}
2 \times 70 + 3 \times 21 + 2 \times 15 \equiv 2 & \pmod 3 \\
2 \times 70 + 3 \times 21 + 2 \times 15 \equiv 3 & \pmod 5 \\
2 \times 70 + 3 \times 21 + 2 \times 15 \equiv 2 & \pmod 7
\end{array}
$$

And the remainders are unchanged if we add a multiple of $3 \times 5 \times 7 = 105$. Therefore the solution to the system of modular equations is

$$ x \equiv 233 \equiv 23 \pmod{105}. $$

Note: How do you apply this method to prove the general Chinese Remainder Theorem?

1.1 Example Questions

Problem 1.1 Find prime factorization of 2018, 2019, 2020, and 2021. (Favorite numbers to be used in math contests.)

Problem 1.2 An eight-digit number $\overline{695xy155}$ is a multiple of 99. Find $10x + y$.

Problem 1.3 Find prime number p such that $5p + 1$ is a perfect cube.

Problem 1.4 Let r be the common remainder when 1405, 2122 and 3795 are divided by $d > 1$. Find $d - r$.

Problem 1.5 Find all positive integers n for which $(n + 1) \mid (n^2 + 1)$.

Problem 1.6 Assume k is a positive integer, and a is a rational number. If a^k is an integer, does a have to be an integer? Justify your answer.

Problem 1.7 Are there any perfect squares in the sequence: $44, 444, 4444, 44444, \ldots$? Justify your answer.

Problem 1.8 Find the remainder when $2^{50} + 41^{65}$ is divided by 7.

Problem 1.9 Find the prime factorization of the number 1000027.

Problem 1.10 Determine whether each of the following conclusion is true. If it is, prove it; if not, give a counterexample.

(a) If $a^2 \equiv b^2 \pmod{m}$, then $a \equiv b \pmod{m}$.

(b) If $a^2 \equiv b^2 \pmod{m}$, then at least one of $a \equiv b \pmod{m}$ or $a \equiv -b \pmod{m}$ is true.

(c) If $a \equiv b \pmod{m}$, then $a^2 \equiv b^2 \pmod{m^2}$.

(d) If $a \equiv b \pmod{2}$, then $a^2 \equiv b^2 \pmod{2^2}$.

(e) If p is an odd prime number, $p \nmid a$, then $a^2 \equiv b^2 \pmod{p}$ if and only if exactly one the following is true: either $a \equiv b \pmod{p}$ or $a \equiv -b \pmod{p}$.

(f) Assume $\gcd(a, m) = 1$, and $k \geq 1$. If $a^k \equiv b^k \pmod{m}$ and $a^{k+1} \equiv b^{k+1} \pmod{m}$, then $a \equiv b \pmod{m}$.

1.2 Practice Questions

Problem 1.11 Show that $70! \equiv 61! \pmod{71}$.

Problem 1.12 Find the prime factorization of the number 1728000001.

Problem 1.13 Find all triples of positive integers (a, b, c) such that $a \equiv b \pmod{c}$, $b \equiv c \pmod{a}$, and $c \equiv a \pmod{b}$.

Problem 1.14 If a, b, c, d are positive integers and $ad - bc = 1$, show that $\dfrac{a+b}{c+d}$ is irreducible.

Problem 1.15 A positive integer n has a units digit 2. If this units digit 2 is moved to the first digit, a new number is formed, and the new number equals $2n$. Find the smallest such number n.

Note: n has more than 15 digits.

Problem 1.16 If $m > 2$, and $a_1, a_2, \ldots, a_k$ is a reduced residue system modulo m, show that $\displaystyle\sum_{i=1}^{k} a_i \equiv 0 \pmod{m}$.

Problem 1.17 Let m be an even number, and let $a_1, a_2, \ldots, a_m$ and $b_1, b_2, \ldots, b_m$ be any two complete residue systems modulo m, show that $a_1 + b_1, a_2 + b_2, \ldots, a_m + b_m$ is *not* a complete residue system modulo m.

Problem 1.18 Assume $m > 2$, show that $0^2, 1^2, 2^2, \ldots, (m-1)^2$ is *not* a complete residue system modulo m.

Problem 1.19 Let k be an odd positive integer, and n be an arbitrary positive integer. Show that

$$(1 + 2 + \cdots + n) \mid (1^k + 2^k + \cdots + n^k)$$

Problem 1.20 Find all six-digit numbers $\overline{abcdef}$, so that when multiplied by $2, 3, 4, 5, 6$, the results are also six-digit numbers with a, b, c, d, e, f as their digits.

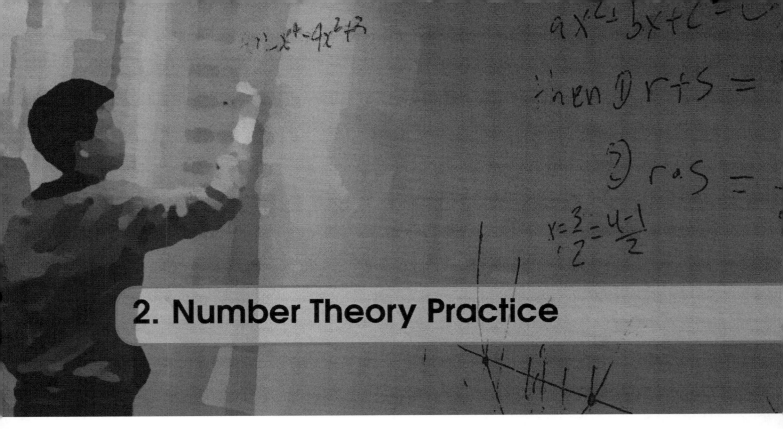

2. Number Theory Practice

In this chapter we do more practice with problems in number theory. There are various common techniques for solving number theory problems.

- Expressing Integers in Different Ways. Common forms of expressions are:
  - Place values: $n = a_k \cdot 10^k + a_{k-1} \cdot 10^{k-1} + \cdots + a_1 \cdot 10 + a_0$;
  - Division with a remainder: $n = mq + r$, where q is the quotient and r is the remainder, and $r < |m|$;
  - Prime factorization: $n = p_1^{e_1} p_2^{e_2} \cdots p_k^{e_k}$;
  - Power of 2 times an odd number: $n = 2^m \cdot t$ where t is an odd number.
- Enumeration: Enumeration method can be described as "smart brute force." Case analysis is one example. Common methods of Enumeration include categorizing based on residue classes, parity, or values.
- Pairing: A lot of problems can be solved by the application of pairing. Legend has it that Gauss used pairing in elementary school to solve the problem $1 + 2 + 3 + \cdots + 100$.
- Parity Analysis: Analyzing the parity (the property of being even or odd) of a quantity can be very useful.

2.1 Example Questions

Problem 2.1 The sum of the first k positive integers is a 3-digit number whose digits are all the same. Find this 3-digit number.

Problem 2.2 Let a, b, c, d, e be five consecutive positive integers. Let

$$m = a + b + c + d + e \text{ and } n = a \cdot b \cdot c \cdot d \cdot e.$$

Also, let the last two digits of $m \cdot n$ be $\overline{xy}$. Find all possible values of x and y.

Problem 2.3 In a mathemagic show, the mathemagician asked Nick (a person he picked from the audience) to (1) think about a three digit number $\overline{abc}$; and (2) write down five numbers (not letting the mathemagician see): $\overline{acb}$, $\overline{bac}$, $\overline{bca}$, $\overline{cab}$, $\overline{cba}$; and (3) add up these five numbers to get N. As soon as Nick said the value of N, the mathemagician announced the original number $\overline{abc}$. If $N = 3194$, what was $\overline{abc}$?

Problem 2.4 Partition the first n positive integers into several non-intersecting subsets, so that none of the subsets contain both m and $2m$ for any m. At least how many subsets should there be?

Problem 2.5 Attach a positive integer N to the right of any positive integer (for example, attaching 8 to the right of 57, we get 578), if the new number is always divisible by N no matter what the other positive integer is, then call N a "magic number". Find all "magic numbers" less than 2000.

Problem 2.6 Ginny has a deck of 100 cards. She starts with the card on top, and do the following: throw away the top card, and put the next top card to the bottom; then throw away the new top card, and put the next top card at the bottom, and so on, until only one card is left. Which card from the original deck is the remaining card?

Problem 2.7 A certain 4-digit number satisfy the following: its tens digit minus 1 equals its units digit; the units digit plus 2 equals the hundreds digit; and if the digits of this 4-digit number is reversed, the new number plus the original number equals 9878. Find the original 4-digit number.

Problem 2.8 Let a, b, c, d be a permutation of the numbers $1, 2, 3, 4$, satisfying $a < b, b > c, c < d$, and $\overline{abcd}$ is a 4-digit number. Find all such 4-digit numbers.

Problem 2.9 Let n be the smallest multiple of 75 that has exactly 75 factors. Find $\dfrac{n}{75}$.

Problem 2.10 What is the largest even number that cannot be written as the sum of two odd composite numbers?

Problem 2.11 Find three prime numbers whose product is five times their sum.

Problem 2.12 From natural numbers $1, 2, 3, \ldots, 1000$, at most how many can be selected such that the sum of any three of the selected numbers is a multiple of 18?

Problem 2.13 A book has 200 pages and is printed double-sided on 96 sheets of paper. Each page has its page number printed at a corner. Luke tore 25 sheets out of the book, and added up all the page numbers printed on these sheets. Is it possible that the sum is 2020?

Problem 2.14 Find the sum of all the digits in the numbers $1, 2, 3, \ldots, 9999999$.

Problem 2.15 Seventy-seven coins are put on the table, showing heads. First turn over all 77 coins. The second step, turn over 76 of them. The third step, turn over 75 of them, and so on. The 77th step, only turn over 1 of the coins. Is it possible to make all 77 coins show tails? If not, explain why. If yes, describe how it can be done.

Problem 2.16 Evaluate:

$$\left\lfloor \frac{199 \times 1}{97} \right\rfloor + \left\lfloor \frac{199 \times 2}{97} \right\rfloor + \cdots + \left\lfloor \frac{199 \times 96}{97} \right\rfloor.$$

Problem 2.17 (Putnam 1989) Let K be the set of all positive integers consisting of alternating digits 1 and 0: $\{1, 101, 10101, 1010101, \ldots\}$. Which elements of K are prime numbers?

2.2   Practice Questions

Problem 2.18 Let p and q be primes. The equation $x^4 - px^3 + q = 0$ has an integer root. Find the values of p and q.

Problem 2.19 The Gauss Middle School has 98 students, each has a unique student number, from 1 through 98. Is it possible to let the students stand in several rows, such that there is a student in each row whose number equals the sum of the numbers of the rest of the students in the same row?

Problem 2.20 A department store distributes 9999 raffle tickets to the customers, each ticket has a 4-digit number from 0001 to 9999. If the sum of the first two digit equals the sum of the last two digits, then the ticket is called a "lucky ticket". For example, ticket number 0945 is a lucky ticket. Show that the sum of all the numbers on the lucky tickets is divisible by 101.

Problem 2.21 Arrange the numbers $1, 2, 3, \ldots, 999$ on a circle, in that order. Start from 1, do the following: skip 1, cross out 2 and 3; skip 4, cross out 5 and 6. Each step skip one number and cross out the next two. Which number is the last one remaining?

Problem 2.22 Form 4-digit numbers with the digits $0, 1, 2, 3, 4$, with no repeating digits within each number (for example, 1023 and 3412). Find the sum of all such 4-digit numbers.

Problem 2.23 Twenty-seven countries send delegations to an international conference, each country has two representatives. Is it possible to arrange the 54 people around a round table, so that between the two people from any country, there are 9 people from other countries?

Problem 2.24 From the numbers $1, 2, 3, \ldots, 999$, cross out the least possible number of numbers so that none of the remaining numbers is the product of two other remaining numbers. Which numbers should be crossed out?

Problem 2.25 A magic coin machine behaves as follows. If you put in a penny, it returns a dime and a nickel. If you put in a nickel, it returns 4 dimes. If you put in a dime, it returns 3 pennies. Becky started with a penny and a nickel, and kept putting coins into the machine and collected the returned coins. Is it possible at some point of time that the number of pennies Becky has is exactly 10 less than the number of dimes?

Problem 2.26 The 3×3 table below contains 9 primes numbers. Define an "operation" as adding the same positive integer to the 3 numbers in one row or one column. Is it possible to change all numbers in the table to the same number after several operations?

2	3	5
13	11	7
17	19	23

Problem 2.27 Let

$$\frac{m}{n} = 1 + \frac{1}{2} + \frac{1}{3} + \cdots + \frac{1}{88}$$

where $\gcd(m, n) = 1$. Show that $89 \mid m$.

Problem 2.28 One integer n equals the sum of 4 distinct fractions of form $\dfrac{m}{m+1}$ (m is a positive integer). Find this integer n, and also find at least one such set of 4 fractions that add up to n. Bonus question: find all such sets of fractions.

Problem 2.29 Ninety-nine students participated in the Planetary Math Olympiad. There are 30 questions in the PMO, and the scoring is as follows. There are 15 base points; add 5 for each correct answer, subtract 1 for each incorrect answer, and add 1 for each unanswered question. If all the scores were added up, is the sum an even number or an odd number?

Problem 2.30 A magic square is a square matrix with the property that the sums of the numbers on each row, column, and diagonal are the same. This sum is called the "magic sum". Is it possible that a 3×3 magic square has a magic sum 1999?

Problem 2.31 A five-digit number N consists of 5 distinct nonzero digits, and N equals the sum of all possible 3-digit numbers made up of 3 of its 5 digits. Find all such 5-digit numbers N.

Problem 2.32 There are the 909 numbers on the board, $1, 2, \ldots, 909$. Each step, erase any two numbers from the board and write their nonnegative difference onto the board, until there is only one number left. Is this last number even or odd?

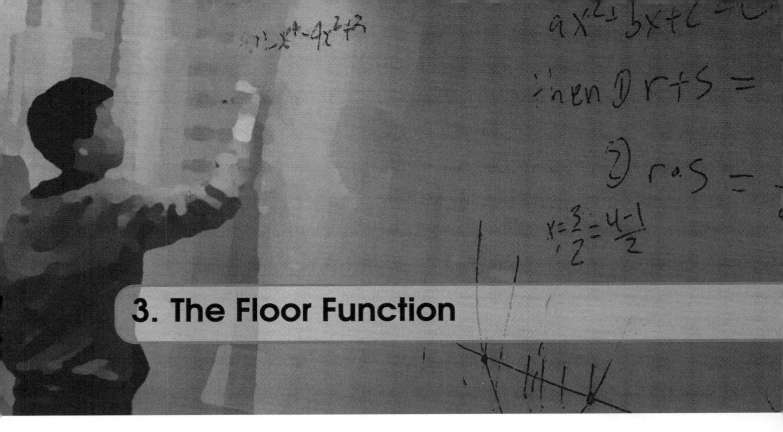

3. The Floor Function

The Floor Function (also called The Integer Part)

Let x be a real number, then denote as $\lfloor x \rfloor$ the largest integer not exceeding x. For example, $\lfloor 2 \rfloor = 2$, $\lfloor \sqrt{3} \rfloor = 1$, $\lfloor -5.5 \rfloor = -6$.

The Fractional Part

For real number x, define $\{x\} = x - \lfloor x \rfloor$. This function is the fractional part of the real number x, therefore $0 \leq \{x\} < 1$. For example, $\{3\} = 0$, $\{9.25\} = 0.25$, $\{-3.14\} = 0.86$.

Properties of $\lfloor x \rfloor$ and $\{x\}$

Let x and y be real numbers:

- $x = \lfloor x \rfloor + \{x\}$;
- $x - 1 < \lfloor x \rfloor \leq x < \lfloor x \rfloor + 1$;
- If $x \leq y$, then $\lfloor x \rfloor \leq \lfloor y \rfloor$;
- Let n be an integer, then $\lfloor x + n \rfloor = \lfloor x \rfloor + n$;
- $\lfloor x + y \rfloor \geq \lfloor x \rfloor + \lfloor y \rfloor$;
- If $\lfloor x \rfloor = \lfloor y \rfloor$, then $|x - y| < 1$;
- If $x \geq 0$ and $y \geq 0$, then $\lfloor xy \rfloor \geq \lfloor x \rfloor \lfloor y \rfloor$;
- If x is an integer, $\lfloor -x \rfloor = -\lfloor x \rfloor$; if x is not an integer, $\lfloor -x \rfloor = -\lfloor x \rfloor - 1$;

- Let n be a positive integer, then $\left\lfloor \dfrac{\lfloor x \rfloor}{n} \right\rfloor = \left\lfloor \dfrac{x}{n} \right\rfloor$;

    Proof. Let $x = nq + r$ where q is an integer and r is a real number satisfying $0 \leq r < n$. Then $\lfloor x \rfloor = nq + \lfloor r \rfloor$, and $\left\lfloor \dfrac{\lfloor x \rfloor}{n} \right\rfloor = \left\lfloor q + \dfrac{\lfloor r \rfloor}{n} \right\rfloor = q$, and $\left\lfloor \dfrac{x}{n} \right\rfloor = \left\lfloor q + \dfrac{r}{n} \right\rfloor = q$. ∎

- In $n!$, the maximum power of a prime number p is

$$\left\lfloor \frac{n}{p} \right\rfloor + \left\lfloor \frac{n}{p^2} \right\rfloor + \left\lfloor \frac{n}{p^3} \right\rfloor + \cdots$$

Example 3.1

Rounding. Let $x > 0$ be a real number. To round x to the nearest integer, we can just calculate $\lfloor x + 0.5 \rfloor$.
   (a) Round 32.54 to the nearest integer: $\lfloor 32.54 + 0.5 \rfloor = \lfloor 33.04 \rfloor = 33$.
   (b) Round 9.48 to the nearest integer: $\lfloor 9.48 + 0.5 \rfloor = \lfloor 9.98 \rfloor = 9$.
Note: This formula is very useful in computer programming.

Example 3.2

How many multiples of 13 are between 100 and 500?
Solution: The number of multiples of 13 between 1 and 500 is $\left\lfloor \dfrac{500}{13} \right\rfloor = 38$; the number of multiples of 13 between 1 and 100 is $\left\lfloor \dfrac{100}{13} \right\rfloor = 7$. So the answer is $38 - 7 = 31$.

Example 3.3

Find the number of zeros at the end of $2017!$.

Solution: Each zero is generated by a pair of factors 2 and 5. Since there are more factors of 2 than factors of 5, the zeros are determined by the maximum power of 5. So the number of zeros is:

$$\left\lfloor \frac{2017}{5} \right\rfloor + \left\lfloor \frac{2017}{5^2} \right\rfloor + \left\lfloor \frac{2017}{5^3} \right\rfloor + \left\lfloor \frac{2017}{5^4} \right\rfloor = 403 + 80 + 16 + 3 = 502.$$

Example 3.4

Solve for x: $\lfloor 3x + 1 \rfloor = 2x - \dfrac{1}{2}$ ♣

Solution: $-3/4$ and $-5/4$. Change variable: $t = 2x - \dfrac{1}{2}$. Thus t must be an integer. And we get $x = \dfrac{t}{2} + \dfrac{1}{4}$, so $3x + 1 = \dfrac{3t}{2} + \dfrac{7}{4}$. Hence $\left\lfloor \dfrac{3t}{2} + \dfrac{7}{4} \right\rfloor = t$. Based on the properties of the floor function, $t \le \dfrac{3t}{2} + \dfrac{7}{4} < t + 1$, solve and get $-\dfrac{7}{2} \le t < -\dfrac{3}{2}$, therefore $t = -2$ or -3. The corresponding values for x are $-\dfrac{3}{4}$ and $-\dfrac{5}{4}$.

Example 3.5

Prove that $\lfloor x \rfloor + \left\lfloor x + \dfrac{1}{2} \right\rfloor = \lfloor 2x \rfloor$ for any real number x. ♣

Solution: Let $n = \lfloor x \rfloor$ and $a = \{x\}$, so $0 \le a < 1$. We prove it in two cases:

Case 1: $0 \le a < \dfrac{1}{2}$. In this case, $0 \le 2a < 1$, and $\left\lfloor x + \dfrac{1}{2} \right\rfloor = n$, and the LHS equals $2n$. The RHS is $\lfloor 2n + 2a \rfloor = 2n$. Therefore both sides are equal.

Case 2: $\dfrac{1}{2} \leq a < 1$. In this case $1 \leq 2a < 2$, therefore $\left\lfloor x + \dfrac{1}{2} \right\rfloor = n+1$, the LHS equals $2n+1$, and the RHS is $\lfloor 2n + 2a \rfloor = 2n+1$, and both sides are still equal.

Combining the two cases, $\lfloor x \rfloor + \left\lfloor x + \dfrac{1}{2} \right\rfloor = \lfloor 2x \rfloor$ for any real number x.   ■

Note: The technique in this example, let $n = \lfloor x \rfloor$ and $a = \{x\}$, is very commonly used in problems involving the Floor function.

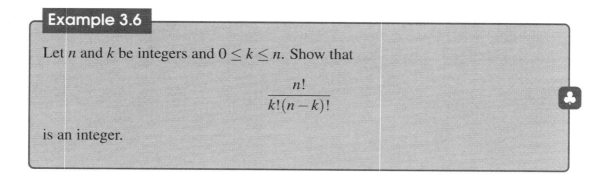

Example 3.6

Let n and k be integers and $0 \leq k \leq n$. Show that

$$\frac{n!}{k!(n-k)!}$$

is an integer.

Proof. We only need to show that for any prime factor p, the maximum power of p in $n!$ is greater than or equal to the sum of the maximum powers of p in $k!$ and $(n-k)!$. Based on the properties mentioned above, the maximum power of p in $n!$ is

$$\left\lfloor \frac{n}{p} \right\rfloor + \left\lfloor \frac{n}{p^2} \right\rfloor + \left\lfloor \frac{n}{p^3} \right\rfloor + \cdots = \sum_{i=1}^{\infty} \left\lfloor \frac{n}{p^i} \right\rfloor.$$

Similarly, the maximum power of p in $k!(n-k)!$ is

$$\sum_{i=1}^{\infty} \left(\left\lfloor \frac{k}{p^i} \right\rfloor + \left\lfloor \frac{n-k}{p^i} \right\rfloor \right).$$

Since

$$\sum_{i=1}^{\infty} \left(\left\lfloor \frac{k}{p^i} \right\rfloor + \left\lfloor \frac{n-k}{p^i} \right\rfloor \right) \leq \sum_{i=1}^{\infty} \left\lfloor \frac{k+n-k}{p^i} \right\rfloor = \sum_{i=1}^{\infty} \left\lfloor \frac{n}{p^i} \right\rfloor,$$

the conclusion follows.   ■

3.1 Example Questions

Problem 3.1 Draw the graph of the function $y = \lfloor x \rfloor$.

Problem 3.2 Evaluate the following:

(a) $\lfloor -\sqrt{2020} \rfloor$.

(b) $\left\lfloor \sqrt{800^2 + 1} + 1 - \sqrt{2} \right\rfloor$.

Problem 3.3 Let $x = \dfrac{1}{3 - \sqrt{7}}$. Find $\lfloor x \rfloor + (1 + \sqrt{7})\{x\}$.

Problem 3.4 Find all integers x that satisfy $\lfloor -1.77x \rfloor = \lfloor -1.77 \rfloor x$.

Problem 3.5 Solve for x: $3x + 5 \lfloor x \rfloor - 50 = 0$.

Problem 3.6 Solve for x: $\left\lfloor \dfrac{5 + 6x}{8} \right\rfloor = \dfrac{15x - 7}{5}$.

Problem 3.7 Solve for x: $\lfloor x \rfloor^3 - 2 \lfloor x^2 \rfloor + \lfloor x \rfloor = 1 - x$.

Problem 3.8 Solve for x: $\lfloor x \rfloor^2 = x \cdot \{x\}$ $(x > 1)$.

Problem 3.9 Solve for x: $\lfloor x^2 \rfloor = \lfloor x \rfloor^2$ for $x \geq 0$.

Problem 3.10 Solve for x: $4x^2 - 40\lfloor x \rfloor + 51 = 0$.

Problem 3.11 Solve for x: $x + \{x\} = 2\lfloor x \rfloor$ $(x \neq 0)$.

Problem 3.12 Solve for x: $x^2 - 4x + 2\lfloor x \rfloor^2 = 0$.

Problem 3.13

(a) Find all real numbers x such that $\{x\} + \left\{\dfrac{1}{x}\right\} = 1$.

(b) Show that any x that satisfies the equation above is irrational.

3.2   Practice Questions

Problem 3.14 (Canada 1981) Are there any real numbers x such that $\lfloor x \rfloor + \lfloor 2x \rfloor + \lfloor 4x \rfloor + \lfloor 8x \rfloor + \lfloor 16x \rfloor + \lfloor 32x \rfloor = 12345$? Justify your answer.

Problem 3.15 If n is a positive integer, find $\left\lfloor \sqrt[3]{n^3 + n^2 + n + 1} \right\rfloor$.

Problem 3.16 Solve for x: $x^3 - 3 = \lfloor x \rfloor$.

Problem 3.17 Find positive integers n such that $\left\lfloor \dfrac{n^2}{4} \right\rfloor$ is prime.

Problem 3.18 Find positive integer n such that $\lfloor \sqrt{n} \rfloor$ is a factor of n.

Problem 3.19 For any positive integer number n and real number x, show that

$$\lfloor x \rfloor + \left\lfloor x + \frac{1}{n} \right\rfloor + \left\lfloor x + \frac{2}{n} \right\rfloor + \cdots + \left\lfloor x + \frac{n-1}{n} \right\rfloor = \lfloor nx \rfloor .$$

Problem 3.20 (Four Countries - Finland etc. 1980) Find the last digit of

$$\left\lfloor (\sqrt{3}+\sqrt{2})^{1980} \right\rfloor .$$

Problem 3.21 Find the last two digits of $\left\lfloor (\sqrt{29}+\sqrt{21})^{1984} \right\rfloor$.

Problem 3.22 Given that $1 \le x < 7$, how many solutions does the following equation have?

$$\{x^2\} = \{x\}^2 .$$

Problem 3.23 Solve the inequality $\lfloor x \rfloor \{x\} < x - 1$.

Problem 3.24 Let $S = 1 + \dfrac{1}{\sqrt{3}} + \dfrac{1}{\sqrt{5}} + \cdots + \dfrac{1}{\sqrt{289}}$. Find the value of $\lfloor S \rfloor$.

Problem 3.25 Evaluate the sum

$$\left\lfloor \frac{101 \times 1}{2017} \right\rfloor + \left\lfloor \frac{101 \times 2}{2017} \right\rfloor + \left\lfloor \frac{101 \times 3}{2017} \right\rfloor + \cdots + \left\lfloor \frac{101 \times 2016}{2017} \right\rfloor .$$

Problem 3.26 In the sequence $\left\lfloor \dfrac{1^2}{1998} \right\rfloor, \left\lfloor \dfrac{2^2}{1998} \right\rfloor, \ldots, \left\lfloor \dfrac{1998^2}{1998} \right\rfloor$, how many distinct values are there?

Problem 3.27 Let $n > 1$ be a positive integer. Show that

$$\lfloor \log_2 n \rfloor + \lfloor \log_3 n \rfloor + \cdots + \lfloor \log_n n \rfloor = \lfloor \sqrt{n} \rfloor + \lfloor \sqrt[3]{n} \rfloor + \cdots + \lfloor \sqrt[n]{n} \rfloor.$$

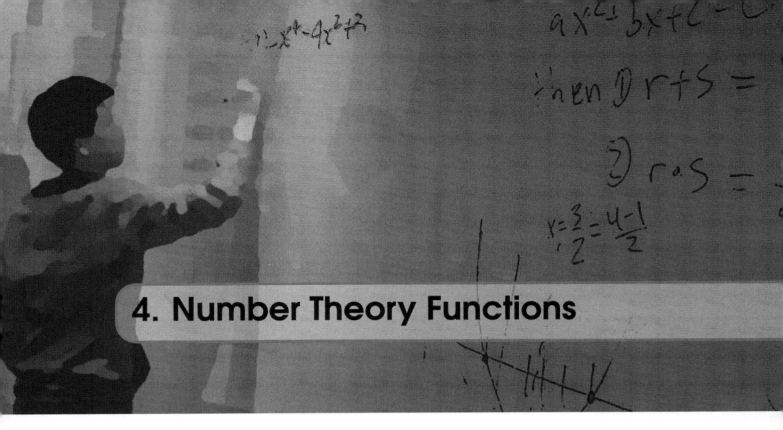

4. Number Theory Functions

Number Theory Functions

If the domain of a function is the set of positive integers, the function is a number theory function, often called an *arithmetic function*.

Any function can be an arithmetic function. For example, the identity function $f(n) = n$ for all $n \in \mathbb{N}$, or the polynomial $f(n) = an^2 + bn + c$, as long as they are defined on the positive integers. Usually, we focus on the functions that are closely related to the number theory properties of the variables; for example, the number of divisors of n, the number of positive integers not exceeding n but relatively prime to n.

Euler's Totient Function

Euler's Totient Function ϕ

Let m be a positive integer, the function $\phi(m)$ is defined as the number of elements in a reduced system of residue classes modulo m. This function is called Euler's Totient Function.

Note: In other words, $\phi(m)$ represents the number of elements in a complete system of residue classes modulo m that are relatively prime to m.

Example 4.1

It is easy to see the following are true.

(i) $\phi(1) = 1$, $\phi(2) = 1$, $\phi(3) = 2$, $\phi(4) = 2$, $\phi(7) = 6$, $\phi(10) = 4$.

(ii) $\phi(p) = p - 1$, where p is a prime number.

(iii) $\phi(p^k) = p^k - p^{k-1} = p^k \left(1 - \dfrac{1}{p} \right)$, where p is prime and k is a positive integer.

Mutiplicative Function

If a function $f(n)$, not identically 0, defined on the positive integers $\mathbb{N}$, satisfies the following property: $f(n_1 n_2) = f(n_1) f(n_2)$ given that $\gcd(n_1, n_2) = 1$, then $f(n)$ is said to be *multiplicative*.

Remark

- Functions such as $f(n) = n$ and $f(n) = n^\alpha$ are clearly multiplicative.
- If $f(n)$ is multiplicative, then $f(1) = 1$.

Theorem 4.1

The Euler's Totient Function $\phi(n)$ is multiplicative. That is, if m and n are positive integers and $\gcd(m, n) = 1$, $\phi(mn) = \phi(m)\phi(n)$.

Theorem 4.2

Let n be a positive integer and its standard prime factorization is $n = p_1^{e_1} p_2^{e_2} \cdots p_k^{e_k}$. Then

$$\phi(n) = (p_1^{e_1} - p_1^{e_1-1})(p_2^{e_2} - p_2^{e_2-1}) \cdots (p_k^{e_k} - p_k^{e_k-1})$$

$$= n \left(1 - \frac{1}{p_1}\right) \left(1 - \frac{1}{p_2}\right) \cdots \left(1 - \frac{1}{p_k}\right).$$

Theorem 4.3

Let m be a positive integer and a be an integer such that $\gcd(a,m) = 1$. If $b_1, b_2, \ldots, b_{\phi(m)}$ is a reduced system of residue classes modulo m, then $ab_1, ab_2, \ldots, ab_{\phi(m)}$ is also a reduced system of residue classes modulo m.

Theorem 4.4 Euler's Theorem

Let m be a positive integer and a be an integer such that $\gcd(a,m) = 1$. Then

$$a^{\phi(m)} \equiv 1 \pmod{m}.$$

More Number Theory Functions

- **Number of divisors**: given positive integer n, denote as $\tau(n)$ the number of positive divisors of n:
$$\tau(n) = \sum_{d|n} 1,$$

Here the sum is over all divisors d of n.
If $n = p_1^{e_1} \cdots p_k^{e_k}$ is the prime factorization of n, then

$$\tau(n) = (e_1 + 1) \cdots (e_k + 1).$$

- **Sum of divisors**: given positive integer n, the sum of its positive divisors is denoted by $\sigma(n)$:

$$\sigma(n) = \sum_{d \mid n} d.$$

If $n = p_1^{e_1} \cdots p_k^{e_k}$ is the prime factorization of n, then

$$
\begin{aligned}
\sigma(n) &= (1 + p_1 + p_1^2 + \cdots + p_1^{e_1})(1 + p_2 + p_2^2 + \cdots + p_2^{e_2}) \cdots \\
&\quad \cdots (1 + p_k + p_k^2 + \cdots + p_k^{e_k}) \\
&= \frac{p_1^{e_1+1} - 1}{p_1 - 1} \cdot \frac{p_2^{e_2+1} - 1}{p_2 - 1} \cdots \frac{p_k^{e_k+1} - 1}{p_k - 1}
\end{aligned}
$$

- Let α be any complex number, define function

$$\sigma_\alpha(n) = \sum_{d \mid n} d^\alpha.$$

The functions $\tau(n)$ and $\sigma(n)$ are special cases of $\sigma_\alpha(n)$:

$$\sigma_0(n) = \tau(n), \quad \sigma_1(n) = \sigma(n).$$

Remark

Functions $\tau(n)$ and $\sigma(n)$ are multiplicative functions.

Completely Mutiplicative Function

If a function $f(n)$, not identically 0, defined on the positive integers $\mathbb{N}$, satisfies the following property: $f(n_1 n_2) = f(n_1) f(n_2)$ for any positive integers n_1 and n_2, ♡ then $f(n)$ is said to be *completely multiplicative*.

Remark

- Again, functions such as $f(n) = n$ and $f(n) = n^\alpha$ are clearly completely multiplicative.

- If $f(n)$ is completely multiplicative, then it is multiplicative. In particular, $f(1) = 1$.
- The functions $\phi(n)$, $\tau(n)$, $\sigma(n)$, and $\sigma_\alpha(n)$ are all multiplicative, but they are not completely multiplicative.

Other Number Theory Functions (enjoy the Greek alphabet soup)

- Number of prime factors: Let $n = p_1^{e_1} \cdots p_k^{e_k}$ be the standard prime factorization of n, define

$$\omega(n) = \begin{cases} k, & n > 1, \\ 0, & n = 1, \end{cases}$$

  i.e. $\omega(n)$ is the number of distinct prime factors of n; also define

$$\Omega(n) = \begin{cases} e_1 + e_2 + \cdots + e_k, & n > 1, \\ 0, & n = 1, \end{cases}$$

  i.e. $\Omega(n)$ is the number of prime factors of n, multiplicity counted. Clearly $\omega(n)$ and $\Omega(n)$ are not multiplicative, but it is easy to see that

$$\omega(mn) = \omega(m) + \omega(n), \text{ if } \gcd(m,n) = 1;$$

  and

$$\Omega(mn) = \Omega(m) + \Omega(n), \forall n \in \mathbb{N}.$$

- Function $\nu(n) = (-1)^{\omega(n)}$ is multiplicative, but not completely multiplicative.
- **The Liouville function**: $\lambda(n) = (-1)^{\Omega(n)}$ is completely multiplicative.
- **The delta function**: $\delta(1) = 1$ and $\delta(n) = 0$ for all $n > 1$.
- **The Möbius function** $\mu(n)$ is defined as follows:

$$\mu(n) = \begin{cases} 0, & \text{If } n \text{ is divisible by a square of a prime;} \\ 1, & \text{If } n = 1; \\ (-1)^r, & \text{If } n \text{ is the product of } r \text{ distinct primes.} \end{cases}$$

Note: It is easy to verify that $\mu(n)$ is multiplicative, but not completely multiplicative. Also, $\delta(n)$ is completely multiplicative.

Dirichlet convolution: If f and g are two functions defined on $\mathbb{N}$, their *Dirichlet convolution* is a function, denoted $f * g$, defined by

$$(f * g)(n) = \sum_{d|n} f(d)g\left(\frac{n}{d}\right)$$

where the sum extends over all positive divisors d of n.

Theorem 4.5

Properties of Dirichlet convolution:
- Commutative: $f * g = g * f$.
- Associative: $(f * g) * h = f * (g * h)$.
- Distributive law also holds: $f * (g + h) = f * g + f * h$.
- The function $\delta(n)$ is the identity function for Dirichlet convolution: $\delta * f = f * \delta = f$.

Möbius transforms

Theorem 4.6 Möbius transforms

if f and g are two functions defined on $\mathbb{N}$, and satisfy

$$g(n) = \sum_{d|n} f(d),$$

then

$$f(n) = \sum_{d|n} \mu(d) g\left(\frac{n}{d}\right),$$

where $\mu(n)$ is the Möbius function.

Möbius transforms and Inverse Möbius transforms

Given two functions f and g, if

$$g(n) = \sum_{d|n} f(d),$$

then function $g(n)$ is the *Möbius transform* of $f(n)$. According to Theorem 4,

$$f(n) = \sum_{d|n} \mu(d) g\left(\frac{n}{d}\right),$$

and $f(n)$ is called the *inverse Möbius transform* of $g(n)$.

Remark

In terms of Dirichlet convolutions,

$$g = f * \mathbf{1}; \qquad f = \mu * g,$$

where $\mathbf{1}(n) = 1$ (for all $n \in \mathbb{N}$) is a constant function.

Theorem 4.7

If $f(n)$ is a multiplicative function, then its Möbius transform, $g(n) = \sum_{d|n} f(d)$, is also multiplicative.

4.1 Example Questions

Problem 4.1 Let $m > 1$ be an integer. What is the sum of all positive integers less than m and relative prime to m?

Problem 4.2 Let m be a positive integer. Find $\sum_{d|m} \phi(d)$. Here the sum is taken over all positive divisors d of m.

Problem 4.3 Find all positive integers n such that $\phi(n) = \frac{1}{2}n$.

Problem 4.4 Simplify: $\sum_{d|n} \frac{1}{d}$.

Problem 4.5 Find the Möbius transform of $\mu(n)$.

Problem 4.6 Find all positive integers n such that $3 \nmid \phi(n)$.

Problem 4.7 Show that for any positive integers m, n,

$$\phi(mn)\phi(\gcd(m,n)) = \gcd(m,n)\phi(m)\phi(n).$$

Problem 4.8 Find all positive integers n such that $\phi(n) = \frac{1}{3}n$.

Problem 4.9 Let m and n be positive integers. Show that

$$\phi(mn) = \gcd(m,n)\phi(\operatorname{lcm}(m,n)).$$

Problem 4.10 Find the smallest positive integer k such that

(a) $\phi(n) = k$ has no solutions for n.

(b) $\phi(n) = k$ has exactly 2 solutions for n.

(c) $\phi(n) = k$ has exactly 3 solutions for n.

4.2   Practice Questions

Problem 4.11

(a) If $f(n)$ is multiplicative, is $\sum_{d|n}(f(d))^3$ also multiplicative? Justify your answer.

(b) If $f(n)$ is multiplicative, is $\left(\sum_{d|n}f(d)\right)^2$ also multiplicative?

(c) Show that

$$\sum_{d|n}(\tau(d))^3 = \left(\sum_{d|n}\tau(d)\right)^2.$$

Problem 4.12 Let n be a positive integer. Find all possible values of

$$\mu(n)\mu(n+1)\mu(n+2)\mu(n+3).$$

Problem 4.13 Find the sum:

$$\sum_{k=1}^{\infty}\mu(k!).$$

Problem 4.14 Let n be a positive integer, find the result of

$$\sum_{d|n} \mu(d)\sigma(d).$$

Problem 4.15 Is the following sum an even number or an odd number?

$$\tau(1) + \tau(2) + \tau(3) + \cdots + \tau(7000)$$

Review Problems

Problem 4.16 Find the last two digits of $n = 7^{7^7}$.

Problem 4.17 Find the prime factorization of the 12-digit number 999999999999.

Problem 4.18 Let a and b be digits, satisfying $\overline{25ab} = 2^5 \cdot a^b$ ($\overline{25ab}$ is a 4-digit number). Find a and b.

Problem 4.19 Let k be a positive odd integer. Does there exist a positive integer n, such that

$$(n+2) \mid (1^k + 2^k + \cdots + n^k)?$$

Problem 4.20 The base-10 representation of the positive integer n is $\overline{13xy45z}$. Given that $792 \mid n$, find x, y, z.

Problem 4.21 Let $f(x)$ be a polynomial with integer coefficients, and m be a positive integer. Given that none of $f(1), f(2), \ldots, f(m)$ is divisible by m, show that $f(x) = 0$ has no integer solutions.

Problem 4.22 Let P be a convex $2n$-sided polygon, and let $A_1, A_2, \ldots, A_{2n}$ be its vertices. Label the sides of P with the integers $1, 2, \ldots, 2n$, not necessarily in order. Also given that M is a point inside P. Label the segments $MA_1, MA_2, \ldots, MA_{2n}$ with the integers $1, 2, \ldots, 2n$ too, not necessarily in order either. For each of the triangles $\triangle MA_1A_2, \triangle MA_2A_3, \ldots, \triangle MA_nA_1$, calculate the sum of the labels on its 3 sides. Is there a way to place the labels so that the sum of the labels for the triangles are all equal?

Problem 4.23 Show that there are infinitely many positive integers a, such that $n^4 + a$ is a composite number for all positive integers n.

Problem 4.24 Let $M = 1 + 2 + 3 + \cdots + 2016$. Find the remainder when 2016! is divided by M.

Problem 4.25 Let $a_1, a_2, \ldots, a_n$ be integers, and

$$a_1 + a_2 + \cdots + a_n = 0, \qquad a_1 a_2 \cdots a_n = n.$$

What is the remainder when n is divided by 4?

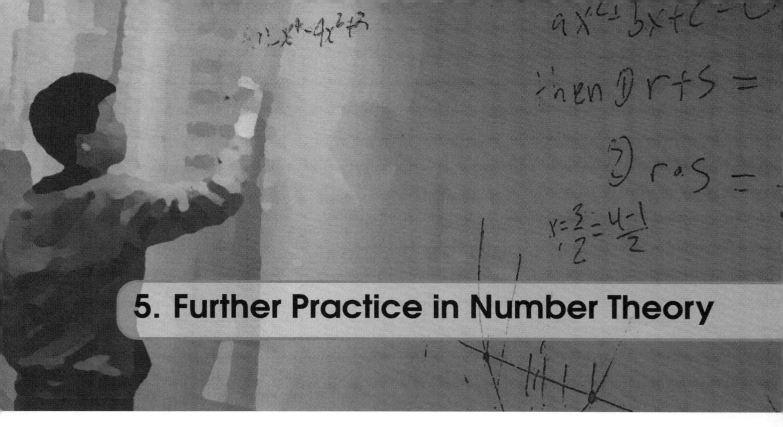

5. Further Practice in Number Theory

In this chapter, we practice with more review problems in number theory.

5.1 Example Questions

Problem 5.1 The equation

$$62 - 63 = 1$$

is obviously false. Can you move only one digit to make the resulting equation true?

Problem 5.2 Can you find ...

(a) A multiple of 1350 that is a perfect cube? (You have *one second* to give an answer)

(b) The smallest positive multiple of 1350 that is a perfect cube?

(c) All positive multiples of 1350 that are perfect cubes?

Problem 5.3 Diophantus was one of the last great Greek mathematicians; he developed his own algebraic notation and is sometimes called "the father of algebra." This riddle about Diophantus' age when he died was carved on his tomb:

> God vouchsafed that he should be a boy for the sixth part of his life; when a twelfth was added, his cheeks acquired a beard; He kindled for him the light of marriage after a seventh, and in the fifth year after his marriage He granted him a son. Alas! late-begotten and miserable child, when he had reached the measure of half his father's life, the chill grave took him. After consoling his grief by this science of numbers for four years, he reached the end of his life.

How long did Diophantus live? (Can you do it without algebra? Can you do it in three seconds?)

Problem 5.4 Attach 3 digits after the number 503 so that the resulting 6-digit integer is a multiple of 7, 9 and 11. Find all such 6-digit integers.

Problem 5.5 What is the smallest positive integer ...

(a) ... that has exactly 10 factors?

(b) ... that has exactly 60 factors?

Problem 5.6 For positive integer $n > 2$, show that there must be a prime number between n and $n!$.

Problem 5.7 Let n be a positive integer greater than 11. Show that n must be the sum of two composite numbers.

Problem 5.8 Let $a_1, a_2, \ldots, a_{64}$ be a rearrangement of $1, 2, 3, \ldots, 64$, and let $b_1 = |a_1 - a_2|, b_2 = |a_3 - a_4|, \ldots, b_{32} = |a_{63} - a_{64}|$. Let $c_1, c_2, \ldots, c_{32}$ be a rearrangement of $b_1, b_2, \ldots, b_{32}$, and then calculate the 16 numbers $|c_1 - c_2|, |c_3 - c_4|, \ldots, |c_{31} - c_{32}|$, and so on, until there is only one number x left. Is x an even number or an odd number, or is it uncertain, depending on the process?

Problem 5.9 Find the remainder when $10^{10} + 10^{10^2} + 10^{10^3} + \cdots + 10^{10^{10}}$ is divided by 7.

Problem 5.10 A perfect square can end with two 4s ($12^2 = 144$). Please answer the following questions.

(a) Can a perfect square end with three 4s? If yes, give an example; if no, prove it.

(b) Can a perfect square end with four 4s? If yes, give an example; if no, prove it.

Problem 5.11 Write down the integers from 1 to 1024 in reverse order: 1024102310 22......54321. Perform the following "operation" on this number: take the first digit 1, multiply by 2 to get 2, add the next digit 0 to get 2, multiply by 2 to get 4, add the next digit 2 to get 6, etc., each time multiply by 2 and add the next digit, until the last digit 1 is added. Now the result is a large integer. Since this result has more than one digit, we perform the same operation as above on it, and keep going until the final result is a one-digit number. What is this one-digit number?

Problem 5.12 Find the last digit of

$$47^{47^{\cdot^{\cdot^{\cdot^{47}}}}},$$

where there are $k(> 1)$ 47s.

Problem 5.13 Find the remainder when 2^{345} is divided by 400.

Problem 5.14 (2018 ZIML Master Round) Find the set of all positive integers n such that

$$1^n + 2^n + 3^n + 4^n + 5^n + 6^n$$

is a multiple of 7.

Problem 5.15 Find the last three digits of

$$1 \times 3 \times 5 \times \cdots \times 1989.$$

Problem 5.16 Old McDonald went to the Market and bought 100 chickens for exactly 100 dollars, among which roosters cost 5 dollars each, hens cost 3 dollars each, and three baby chicks cost 1 dollar. How many chickens of each type did he buy? Find all possible solutions.

Problem 5.17 Show that 1599 is not the sum of 14 perfect 4th powers.

Problem 5.18 Find the maximum positive integer n, such that $3^n \mid 2^{3^m} + 1$ for every positive integer m.

5.2 Practice Questions

Problem 5.19 The positive integers a, b, c and d are all divisible by the positive integer $ab - cd$. Find all possible values of $ab - cd$.

Problem 5.20 Find all positive integer n such that $n^2 + 5n + 16$ is divisible by 169.

Problem 5.21 Let a, b, c, d be four integers, show that

$$12 \mid (b-a)(c-a)(d-a)(d-c)(d-b)(c-b).$$

Problem 5.22 The sum of a set of positive integers is divisible by 30. Show that the sum of the fifth powers of the same set of positive integers is also divisible by 30.

Problem 5.23 For each positive odd integer $n < 10000$, denote as $f(n)$ the number formed by the last four digits of n^9. The set A consists of all odd integers $n < 10000$ such that $f(n) > n$ and set B consists of all odd integers such that $f(n) < n$. Which set contains more elements, A or B?

Problem 5.24 Each raffle ticket has a 6-digit number (could begin with 0). If the sum of the first 3 digits of a number equals the sum of the last 3 digits, then this number is called a "lucky number". Show that sum of all lucky numbers is divisible by 13.

Problem 5.25 In triangle ABC, $AB = 33$, $AC = 21$, and $BC = m$, where m is a positive integer. If there is a point D on $\overline{AB}$ and a point E on $\overline{AC}$ such that $AD = DE = EC = n$, where n is a positive integer, find m and n.

Problem 5.26 Let $a_1 = 1, a_2 = 3$, and for all positive integer n,

$$a_{n+2} = (n+3)a_{n+1} - (n+2)a_n.$$

Find all n such that $11 \mid a_n$.

Problem 5.27 Let a, b, c, d be distinct integers such that the equation $(x-a)(x-b)(x-c)(x-d) - 4 = 0$ has an integer root. Find this integer root in terms of a, b, c, d.

Problem 5.28 Find all ordered pairs of positive integers (m, n), such that the polynomial $1 + x^n + x^{2n} + \cdots + x^{mn}$ is divisible by $1 + x + x^2 + \cdots + x^m$.

Problem 5.29 Find the maximum positive integer n such that $\dfrac{(n-2)^2(n+1)}{2n-1}$ is an integer.

Problem 5.30 Find four integers a, b, c, d such that the value $ax^3 + bx^2 + cx + d$ is 1 when $x = 19$, and the value is 2 when $x = 62$.

Problem 5.31 Show that for any integers x and y, the following

$$x^5 + 3x^4y - 5x^3y^2 - 15x^2y^3 + 4xy^4 + 12y^5$$

can never have a value of 33.

Problem 5.32 Find the minimum positive integer m such that $2^{1990} \mid (1989^m - 1)$.

Problem 5.33 Do there exist 100 distinct positive integers, such that the product of any 5 of them is divisible by the sum of the same 5 integers? If yes, find these integers; if no, explain why.

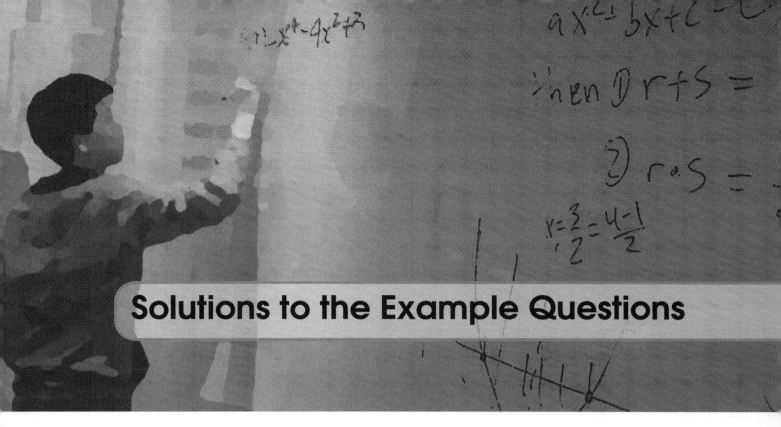

Solutions to the Example Questions

In the sections below you will find solutions to all of the Example Questions contained in this book.

Practice questions are meant to be used for homework, so their answers and solutions are not included. Teachers or math coaches may contact Areteem at info@areteem.org for answer keys and options for purchasing a Teachers' Edition of the course.

1   Solutions to Chapter 1 Examples

Problem 1.1 Find prime factorization of 2018, 2019, 2020, and 2021. (Favorite numbers to be used in math contests.)

Answer

$2018 = 2 \cdot 1009$, $2019 = 3 \cdot 673$, $2020 = 2^2 \cdot 5 \cdot 101$, $2021 = 43 \cdot 47$.

Problem 1.2 An eight-digit number $\overline{695xy155}$ is a multiple of 99. Find $10x + y$.

Answer

32

Solution

For divisibility of 9, add up the digits: $6 + 9 + 5 + x + y + 1 + 5 + 5 = x + y + 31$ is a multiple of 9, so $x + y \equiv 5 \pmod 9$. For divisibility of 11, find alternate sum of the digits: $6 - 9 + 5 - x + y - 1 + 5 - 5 = 1 - x + y$ is a multiple of 11, so $x - y \equiv 1 \pmod{11}$. There is only one pair of digits x and y satisfying both requirements: $x = 3, y = 2$.

Problem 1.3 Find prime number p such that $5p + 1$ is a perfect cube.

Answer

43

Solution

Let $n^3 = 5p + 1$, then $5p = n^3 - 1 = (n - 1)(n^2 + n + 1)$. Either $n - 1 = 5$ or $n^2 + n + 1 = 5$, and the latter doesn't give integer solutions for n. If $n - 1 = 5$, $n = 6$, and $p = n^2 + n + 1 = 43$.

Problem 1.4 Let r be the common remainder when 1405, 2122 and 3795 are divided by $d > 1$. Find $d - r$.

Answer

29

Solution

By the division algorithm there are integers q_1, q_2, q_3 with

$$1405 = dq_1 + r, \quad 2122 = dq_2 + r, \quad 3795 = dq_3 + r.$$

Subtracting we get

$$1673 = d(q_3 - q_2), \quad 717 = d(q_2 - q_1).$$

Notice that d is a common divisor of 1673 and 717. As

$$1673 = 7 \cdot 239, \quad 717 = 3 \cdot 239,$$

we see that 239 is the common divisor greater than 1 so $d = 239$. Since

$$1405 = 5 \cdot 239 + 210,$$

we deduce that $r = 210$. Finally,

$$d - r = 239 - 210 = 29.$$

Problem 1.5 Find all positive integers n for which $(n + 1) \mid (n^2 + 1)$.

Answer

1

Solution

Since $n^2 + 1 = n^2 - 1 + 2 = (n + 1)(n - 1) + 2$, we have $(n + 1) \mid 2$. So $n = 1$ is the only such positive integer.

Problem 1.6 Assume k is a positive integer, and a is a rational number. If a^k is an integer, does a have to be an integer? Justify your answer.

Answer

Yes

Solution

Let $a = \dfrac{p}{q}$ where $q \geq 1$ and $\gcd(p, q) = 1$. If $\left(\dfrac{p}{q}\right)^k = c$ is an integer, then $cq^k = p^k$, so $q \mid p^k$. Since $\gcd(p, q) = 1$, we get $q \mid p$, thus $q = 1$, which is what we need.

Problem 1.7 Are there any perfect squares in the sequence: $44, 444, 4444, 44444, \ldots$? Justify your answer.

Answer

No

Solution

These numbers are all $4 \times 11 \cdots 1$. 4 is a perfect square, but $11 \cdots 1$ with at least 2 digits are not perfect squares because they are all 3 (mod 4).

Problem 1.8 Find the remainder when $2^{50} + 41^{65}$ is divided by 7.

Answer

3

Solution

By Fermat's Little Theorem, $2^6 \equiv 1 \pmod 7$, so $2^{50} \equiv 2^2 \cdot (2^6)^8 \equiv 4 \pmod 7$. Also, $41 \equiv -1 \pmod 7$, so $41^{65} \equiv -1 \pmod 7$. Therefore $2^{50} + 41^{65} \equiv 4 - 1 \equiv 3 \pmod 7$.

Problem 1.9 Find the prime factorization of the number 1000027.

Answer

$7 \times 19 \times 73 \times 103$

Solution

$$
\begin{aligned}
1000027 &= 100^3 + 3^3 \\
&= (100 + 3)(100^2 - 3 \cdot 100 + 3^2) \\
&= (103)(100^2 + 2 \cdot 3 \cdot 100 + 3^2 - 3 \cdot 3 \cdot 100) \\
&= (103)((100 + 3)^2 - 30^2) \\
&= (103)(103 + 30)(103 - 30) \\
&= (103)(133)(73) \\
&= 103 \times 7 \times 19 \times 73.
\end{aligned}
$$

Problem 1.10 Determine whether each of the following conclusion is true. If it is, prove it; if not, give a counterexample.

(a) If $a^2 \equiv b^2 \pmod{m}$, then $a \equiv b \pmod{m}$.

Answer

False

Solution

$1^2 \equiv 3^2 \pmod{4}$ and yet $1 \not\equiv 3 \pmod{4}$.

(b) If $a^2 \equiv b^2 \pmod{m}$, then at least one of $a \equiv b \pmod{m}$ or $a \equiv -b \pmod{m}$ is true.

Answer

False

Solution

$1^2 \equiv 3^2 \pmod{8}$ but $1 \not\equiv 3 \pmod{8}$ and $1 \not\equiv -3 \pmod{8}$.

(c) If $a \equiv b \pmod{m}$, then $a^2 \equiv b^2 \pmod{m^2}$.

Answer

False

Solution

$3 \equiv -1 \pmod{4}$, but $9 \not\equiv 1 \pmod{16}$.

(d) If $a \equiv b \pmod{2}$, then $a^2 \equiv b^2 \pmod{2^2}$.

Answer

True

Solution

$a \equiv b \pmod{2}$ means a and b are both even or both odd, so $a + b$ and $a - b$ are both even, thus $a^2 - b^2 = (a+b)(a-b)$ is a multiple of 4, in other words, $a^2 \equiv b^2 \pmod{2^2}$.

(e) If p is an odd prime number, $p \nmid a$, then $a^2 \equiv b^2 \pmod{p}$ if and only if exactly one the following is true: either $a \equiv b \pmod{p}$ or $a \equiv -b \pmod{p}$.

Answer

True

Solution

"If": assume either $a \equiv b \pmod{p}$ or $a \equiv -b \pmod{p}$ (exactly one is true), then it is obvious that $a^2 \equiv b^2 \pmod{p}$.

"Only if": assume $a^2 \equiv b^2 \pmod{p}$. Thus $p \mid (a-b)(a+b)$, and we conclude that either $p \mid (a-b)$ or $p \mid (a+b)$ is true. We show that they cannot be both true. If they were, then $p \mid (a-b+a+b)$ i.e. $p \mid (2a)$, contradicting the fact that p is an odd prime and $p \nmid a$.

(f) Assume $\gcd(a,m) = 1$, and $k \geq 1$. If $a^k \equiv b^k \pmod{m}$ and $a^{k+1} \equiv b^{k+1} \pmod{m}$, then $a \equiv b \pmod{m}$.

Answer

True

Solution

Let c be the inverse of a^k modulo m, then $1 \equiv ca^k \equiv cb^k \pmod{m}$. So $a \equiv ca^{k+1} \equiv cb^{k+1} \equiv b \pmod{m}$.

2   Solutions to Chapter 2 Examples

Problem 2.1 The sum of the first k positive integers is a 3-digit number whose digits are all the same. Find this 3-digit number.

Answer

666

Solution

Let the three digit number be $\overline{ddd} = 100d + 10d + d = 111d$, where d is a digit. The sum of the first k positive integers is

$$1 + 2 + 3 + \cdots + k = \frac{k(k+1)}{2},$$

so

$$111d = \frac{k(k+1)}{2},$$

then

$$k(k+1) = 222d = 37 \times 6d.$$

This means either k or $k+1$ is a multiple of 37.

If $k = 37m$ ($m \geq 1$), then $37m(37m + 1) = 37 \times 6d$, so $m(37m+1) = 6d$. Since $d \leq 9$, $6d \leq 54$, therefore the only possible value for m is 1, but then $37 + 1 = 6d$ has no integer solution for d.

If $k + 1 = 37m$ ($m \geq 1$), then $(37m - 1)37m = 37 \times 6d$, so $m(37m - 1) = 6d$. Again, $m = 1$, and $36 = 6d$, thus $d = 6$.

Therefore, the 3-digit number is 666.

Problem 2.2 Let a, b, c, d, e be five consecutive positive integers. Let

$$m = a + b + c + d + e \text{ and } n = a \cdot b \cdot c \cdot d \cdot e.$$

Also, let the last two digits of $m \cdot n$ be $\overline{xy}$. Find all possible values of x and y.

Answer

0 and 0

Solution

The sum of five consecutive positive integers is a multiple of 5, so $m = 5k$. Also, there must be at least one multiple of 5, one multiple of 3, and two multiples of 2 among the five numbers, therefore n is a multiple of $2^2 \times 3 \times 5 = 60$. Let $n = 60t$, then $m \cdot n = 5k \cdot 60t = 300kt$. Therefore the last two digits of $m \cdot n$ are both 0.

Problem 2.3 In a mathemagic show, the mathemagician asked Nick (a person he picked from the audience) to (1) think about a three digit number $\overline{abc}$; and (2) write down five numbers (not letting the mathemagician see): $\overline{acb}$, $\overline{bac}$, $\overline{bca}$, $\overline{cab}$, $\overline{cba}$; and (3) add up these five numbers to get N. As soon as Nick said the value of N, the mathemagician announced the original number $\overline{abc}$. If $N = 3194$, what was $\overline{abc}$?

Answer

358

Solution

Adding up all the 3-digit numbers that are permutations of $\overline{abc}$, the sum is $222(a + b + c) = 3194 + \overline{abc}$. The smallest multiple of 222 greater than 3194 is $222 \times 15 = 3330$, and $3330 - 3194 = 136$. However, the sum of digits of 136 is 10, not 15. Thus we look at the next multiple, $222 \times 16 = 3552$, and $3552 - 3194 = 358$, and the sum of digits of 358 is exactly 16. So the original number is 358.

Problem 2.4 Partition the first n positive integers into several non-intersecting subsets, so that none of the subsets contain both m and $2m$ for any m. At least how many subsets should there be?

Answer

2

Solution

Write everything as $2^k a$ where a is an odd number. One set contains all the numbers with even k, one contains odd k.

Problem 2.5 Attach a positive integer N to the right of any positive integer (for example, attaching 8 to the right of 57, we get 578), if the new number is always divisible by N no matter what the other positive integer is, then call N a "magic number". Find all "magic numbers" less than 2000.

Answer

1, 2, 5, 10, 20, 25, 50, 100, 125, 200, 250, 500, 1000, 1250

Solution

Suppose a "magic number" N has k digits. Let x be an arbitrary positive integer, then $10^k \cdot x + N = m \cdot N$. Thus $(m-1)N = 10^k \cdot x$. Because x is arbitrary, this means N is a factor of 10^k. Now we categorize according to the number of digits.

For $k = 1$, $N \mid 10$, so $N = 2$ or 5.

For $k = 2$, $N \mid 100$, so $N = 10, 20, 25, 50$.

For $k = 3$, $N \mid 1000$, so $N = 100, 125, 200, 250, 500$.

For $k = 4$, $N \mid 10000$, so $N = 1000, 1250$ for $N < 2000$.

Thus the numbers are: 1, 2, 5, 10, 20, 25, 50, 100, 125, 200, 250, 500, 1000, 1250, a total of 14 numbers.

Problem 2.6 Ginny has a deck of 100 cards. She starts with the card on top, and do the following: throw away the top card, and put the next top card to the bottom; then throw away the new top card, and put the next top card at the bottom, and so on, until only one card is left. Which card from the original deck is the remaining card?

Answer

The 72nd card

Solution

Start with 1 card and try 2 cards, 3 cards, and so on, and work through the process each time to see which card remains at the end. The pattern is as follows: If $N = 2^n$, then the remaining card is the last one, the 2^n-th. If $N = 2^n + m(m < 2^n)$, then the remaining card is the $2m$-th.

Second Solution:

If there were 64 cards to begin with, the bottom card will be the one that ultimately remains. From 100 cards, we shall find out which card is the bottom card after $100 - 64 = 36$ cards are thrown away.

Every step, one top card is thrown away and the next card is moved to the bottom. So after 36 steps, the 72nd card is at the bottom and there are 64 cards remaining in the deck. Therefore the remaining card is the 72nd card.

Problem 2.7 A certain 4-digit number satisfy the following: its tens digit minus 1 equals its units digit; the units digit plus 2 equals the hundreds digit; and if the digits of this 4-digit number is reversed, the new number plus the original number equals 9878. Find the original 4-digit number.

Answer

1987

Solution

Assume the tens digit is x, then the units digit is $x - 1$ and the hundreds digit is $x + 1$. Let the thousands digit be y. So

$$1000y + 100(x+1) + 10x + (x-1) + 1000(x-1) + 100x + 10(x+1) + y = 9878.$$

Simplifying,
$$1001y + 1221x = 10769.$$
From the last digit, we see that $x + y = 9$, so $y = 9 - x$, and then

$$220x = 1760,$$

so $x = 8$. Hence, $y = 1$, and the 4-digit number is 1987.

Problem 2.8 Let a, b, c, d be a permutation of the numbers $1, 2, 3, 4$, satisfying $a < b, b > c, c < d$, and $\overline{abcd}$ is a 4-digit number. Find all such 4-digit numbers.

Answer

$1324, 1423, 2314, 2413, 3412$

Solution

The value of b can be 3 or 4. Using this fact we can work out all the 5 possible numbers by brute force.

Problem 2.9 Let n be the smallest multiple of 75 that has exactly 75 factors. Find $\dfrac{n}{75}$.

Answer

432

Solution

Since $75 = 3 \times 5^2$, 3 and 5 are prime factors of the number n. To have exactly 75 factors, the prime factorization of n can be $p^4 \cdot 3^4 \cdot 5^2$ (where p is a prime), or $3^{14} \cdot 5^4$, or $3^{24} \cdot 5^2$, or 3^{75}, or these prime factors with different arrangements of exponents. To get the smallest such numbers, we select $p = 2$ and $n = 2^4 \cdot 3^4 \cdot 5^2$ is the smallest among all the possibilities above. Therefore,

$$\frac{n}{75} = 2^4 \cdot 3^3 = 16 \cdot 27 = 432.$$

Problem 2.10 What is the largest even number that cannot be written as the sum of two odd composite numbers?

Answer

38

Solution

The odd composite numbers are 9, 15, 21, 25, 27, 33, 35, ... It is easy to see that 38 cannot be written as a sum of two odd composite numbers. To show it is the largest, consider the following:

$$40 = 25 + 15, \quad 42 = 27 + 15, \quad 44 = 35 + 9.$$

then for any $k \geq 0$,

$$\begin{aligned}
40 + 6k &= 25 + 3(5 + 2k), \\
40 + 6k + 2 &= 27 + 3(5 + 2k), \\
40 + 6k + 4 &= 35 + 3(3 + 2k),
\end{aligned}$$

which means all even numbers greater than 38 can be written as the sum of two odd composite numbers. Therefore, the answer is 38.

Problem 2.11 Find three prime numbers whose product is five times their sum.

Answer

2, 5, 7

Solution

One of the primes must be 5, so let the numbers be $5, p, q$, and

$$5pq = 5(5 + p + q)$$

Thus

$$pq = 5 + p + q$$

Hence

$$pq - p - q + 1 = 6$$
$$(p - 1)(q - 1) = 6$$

There are two choices: $p - 1 = 1, q - 1 = 6$, or $p - 1 = 2, q - 1 = 3$. Only the first one give prime number solutions, so $p = 2, q = 7$.

Problem 2.12 From natural numbers $1, 2, 3, \ldots, 1000$, at most how many can be selected such that the sum of any three of the selected numbers is a multiple of 18?

Answer

56

Solution

Any two of the selected numbers have the same remainder mod 18, and the remainder has to be 0, 6, or 12. There are 56 numbers with remainder 6: $6, 24, 42, \ldots, 996$. The other two remainders both have 55 numbers.

Problem 2.13 A book has 200 pages and is printed double-sided on 96 sheets of paper. Each page has its page number printed at a corner. Luke tore 25 sheets out of the book, and added up all the page numbers printed on these sheets. Is it possible that the sum is 2020?

Answer

No

Solution

The two page numbers on the same sheet of a book must be one even and one odd, with the sum an odd number. The sum of 25 odd numbers is odd, and cannot be 2020.

Problem 2.14 Find the sum of all the digits in the numbers $1, 2, 3, \ldots, 9999999$.

Answer

315000000

Solution

Add 0 to this sum, which does not affect the result.

Pair up 0 and 9999999, 1 and 9999998, etc. Each pair has sum of digits 63, and there are 5000000 pairs. Therefore the total is

$$63 \times 5000000 = 315000000.$$

Problem 2.15 Seventy-seven coins are put on the table, showing heads. First turn over all 77 coins. The second step, turn over 76 of them. The third step, turn over 75 of them, and so on. The 77th step, only turn over 1 of the coins. Is it possible to make all 77 coins show tails? If not, explain why. If yes, describe how it can be done.

Answer

Yes

Solution

In the first step, every coin is turned over. Since the second step turns over 76 coins and the 77th step turns over just 1 coin, paired up these two steps so all coins are included in these steps (for example, turn over the first 76 coins in the second step and turn over the last coin in the 77th step). Similarly, pair up the 3rd step and the 76th step to turn over all coins, and so on, until the 39th step and the 40th step are similarly paired up.

For each pair, all coins are turned. Therefore, including the first step, each coin is turned 39 times, so all are tails at the end.

Problem 2.16 Evaluate:

$$\left\lfloor \frac{199 \times 1}{97} \right\rfloor + \left\lfloor \frac{199 \times 2}{97} \right\rfloor + \cdots + \left\lfloor \frac{199 \times 96}{97} \right\rfloor.$$

Answer

9504

Solution

Pair up $\left\lfloor \dfrac{199 \times 1}{97} \right\rfloor$ and $\left\lfloor \dfrac{199 \times 96}{97} \right\rfloor$. Since

$$\frac{199 \times 1}{97} + \frac{199 \times 96}{97} = 199,$$

and neither fraction is an integer, we get

$$\left\lfloor \frac{199 \times 1}{97} \right\rfloor + \left\lfloor \frac{199 \times 96}{97} \right\rfloor = 198.$$

Similarly,

$$\left\lfloor \frac{199 \times k}{97} \right\rfloor + \left\lfloor \frac{199 \times (97 - k)}{97} \right\rfloor = 198$$

for $1 \leq k \leq 48$, therefore

$$\left\lfloor \frac{199 \times 1}{97} \right\rfloor + \left\lfloor \frac{199 \times 2}{97} \right\rfloor + \cdots + \left\lfloor \frac{199 \times 96}{97} \right\rfloor = 198 \times 48 = 9504.$$

Problem 2.17 (Putnam 1989) Let K be the set of all positive integers consisting of alternating digits 1 and 0: $\{1, 101, 10101, 1010101, \ldots\}$. Which elements of K are prime numbers?

Answer

Only 101

Solution

1 is not a prime; 101 is prime. If there are more than 3 digits, the numbers can be expressed as

$$\frac{10^{2n+2} - 1}{10^2 - 1} = \frac{10^{n+1} - 1}{10 - 1} \cdot \frac{10^{n+1} + 1}{10 + 1} = \frac{111 \cdots 1 \cdot (10^{n+1} + 1)}{10 + 1}.$$

Either n is even or odd, the top is divisible by 11, and the resulting integer is expressed as a product of two integers greater than 1.

Therefore 101 is the only prime number in the set.

3   Solutions to Chapter 3 Examples

Problem 3.1 Draw the graph of the function $y = \lfloor x \rfloor$.

Solution

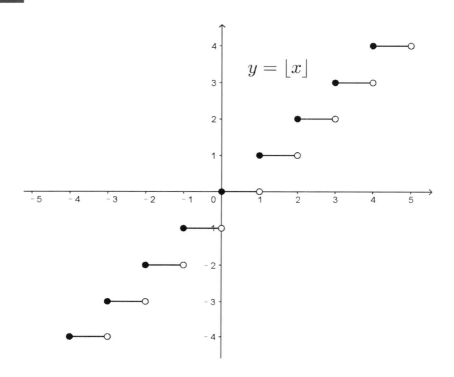

$$y = \lfloor x \rfloor$$

Problem 3.2 Evaluate the following:

(a) $\lfloor -\sqrt{2020} \rfloor$.

Answer

-45

Solution

Since $44^2 = 1936 < 2020 < 2025 = 45^2$, we get $44 < \sqrt{2020} < 45$, and so $-45 < -\sqrt{2020} < -44$, thus $\lfloor -\sqrt{2020} \rfloor = -45$.

(b) $\lfloor \sqrt{800^2 + 1} + 1 - \sqrt{2} \rfloor$.

Answer

799

Solution

Since $800^2 < 800^2 + 1 < 801^2$, we get $\left\lfloor \sqrt{800^2 + 1} \right\rfloor = 800$. Also it appears that the fractional part, $\sqrt{800^2 + 1} - 800$, is smaller than $\sqrt{2} - 1$ so consequently the final answer would be 799, but we need to verify it rigorously. To compare those two values, we rationalize the numerators: $\sqrt{800^2 + 1} - 800 = \dfrac{1}{\sqrt{800^2 + 1} + 800}$, and $\sqrt{2} - 1 = \dfrac{1}{\sqrt{2} + 1}$. The former denominator is way larger than the latter, so we conclude that $\sqrt{800^2 + 1} - 800 < \sqrt{2} - 1$, and therefore $799 < \sqrt{800^2 + 1} + 1 - \sqrt{2} < 800$.

Problem 3.3 Let $x = \dfrac{1}{3 - \sqrt{7}}$. Find $\lfloor x \rfloor + (1 + \sqrt{7})\{x\}$.

Answer

5

Solution

$x = \dfrac{1}{3 - \sqrt{7}} = \dfrac{3 + \sqrt{7}}{2}$, and $2 < \sqrt{7} < 3$, so $2 < x < 3$, thus

$$\lfloor x \rfloor = 2, \quad \{x\} = x - 2 = \frac{\sqrt{7} - 1}{2}.$$

Therefore

$$\lfloor x \rfloor + (1 + \sqrt{7})\{x\} = 2 + (1 + \sqrt{7}) \cdot \frac{\sqrt{7} - 1}{2} = 2 + 3 = 5.$$

Problem 3.4 Find all integers x that satisfy $\lfloor -1.77x \rfloor = \lfloor -1.77 \rfloor x$.

Answer

$0, 1, 2, 3, 4$

Solution

Since $\lfloor -1.77 \rfloor = -2$, the equation can be simplified to $\lfloor -1.77x \rfloor = -2x$. Also,

$$\{-1.77x\} = -1.77x - \lfloor -1.77x \rfloor,$$

which means

$$\lfloor -1.77x \rfloor = -1.77x - \{-1.77x\},$$

so

$$-1.77x - \{-1.77x\} = -2x,$$

then

$$0.23x = \{-1.77x\}.$$

From

$$0 \le \{-1.77x\} < 1,$$

we get

$$0 \le 0.23x < 1,$$

so

$$0 \le x < \frac{100}{23}.$$

Thus the possible values for x are $0, 1, 2, 3, 4$.

Problem 3.5 Solve for x: $3x + 5\lfloor x \rfloor - 50 = 0$.

Answer

$20/3$

Solution

Following the examples, let $t = \lfloor x \rfloor$, which is an integer. So $3x + 5t = 50$, and then $x = \dfrac{50 - 5t}{3}$. By definition, $t \le \dfrac{50 - 5t}{3} < t + 1$, thus $3t \le 50 - 5t < 3t + 3$, and $0 \le 50 - 8t < 3$, so $-50 \le -8t < -47$, hence $6.25 \ge t > 5.875$. The only possible value for t is 6, and therefore $x = 20/3$.

Problem 3.6 Solve for x: $\left\lfloor \dfrac{5 + 6x}{8} \right\rfloor = \dfrac{15x - 7}{5}$.

Answer

$7/15$ and $4/5$

Solution

Let $t = \left\lfloor \dfrac{5+6x}{8} \right\rfloor$. Then $x = \dfrac{5t+7}{15}$. By definition we have $t \le \dfrac{5+6x}{8} < t+1$, thus $t \le \dfrac{10t+39}{40} < t+1$. Solving for t, $0 \le \dfrac{39-30t}{40} < 1$, and get $-\dfrac{1}{30} < t \le \dfrac{13}{10}$. So $t = 0$ or 1, and the corresponding x values are $7/15$ and $4/5$.

Problem 3.7 Solve for x: $\lfloor x \rfloor^3 - 2\lfloor x^2 \rfloor + \lfloor x \rfloor = 1 - x$.

Answer

1

Solution

x must be an integer, so we can get rid of the floor functions: $x^3 - 2x^2 + x = 1 - x$, which is $x^3 - 2x^2 + 2x - 1 = 0$, so $(x-1)(x^2 - x + 1) = 0$. The only real solution is $x = 1$.

Problem 3.8 Solve for x: $\lfloor x \rfloor^2 = x \cdot \{x\}$ $(x > 1)$.

Answer

$(1 + \sqrt{5})/2$

Solution

To simplify the equation, let $n = \lfloor x \rfloor$ and $a = \{x\}$. Thus $x = n + a$, and the equation becomes $n^2 = (n+a)a$. From the requirement of the problem $(x > 1)$, $n \ge 1$, and $a \ne 0$ (otherwise n would have to be 0 too). This equation can be converted to

$$\left(\frac{a}{n}\right)^2 + \left(\frac{a}{n}\right) - 1 = 0,$$

as a quadratic equation in $\dfrac{a}{n}$. Using quadratic formula, $\dfrac{a}{n} = \dfrac{-1+\sqrt{5}}{2}$ (throwing away the negative root). Because $a < 1$ and $-1 + \sqrt{5} > 1$, n cannot be 2 or greater. So $n = 1$, and $a = \dfrac{-1+\sqrt{5}}{2}$, and so $x = \dfrac{1+\sqrt{5}}{2}$.

Problem 3.9 Solve for x: $\lfloor x^2 \rfloor = \lfloor x \rfloor^2$ for $x \ge 0$.

Answer

$n \le x < \sqrt{n^2 + 1}$ for $n = 0, 1, 2, \ldots$

Solution

Let $n = \lfloor x \rfloor$, then $\lfloor x^2 \rfloor = n^2$, so $n^2 \le x^2 < n^2 + 1$, hence $n \le x < \sqrt{n^2 + 1}$. Here n can be any nonnegative integer.

Problem 3.10 Solve for x: $4x^2 - 40 \lfloor x \rfloor + 51 = 0$.

Answer

$\sqrt{29}/2, \sqrt{189}/2, \sqrt{229}/2, \sqrt{269}/2$

Solution

Since $40 \lfloor x \rfloor$ is even, $4x^2$ must be odd (remember x is not an integer).

Let $4x^2 = 2k + 1$, then $x = \dfrac{\sqrt{2k+1}}{2}$, thus

$$\left\lfloor \frac{\sqrt{2k+1}}{2} \right\rfloor = \frac{k + 26}{20}.$$

Since the LHS is an integer, the integer k must be of the form $20m + 14$. Also,

$$\frac{k + 26}{20} \le \frac{\sqrt{2k+1}}{2} < \frac{k + 26}{20} + 1.$$

Simplify and get two inequalities:

$$k^2 - 148k + 4 \times 144 \le 0,$$

and

$$k^2 - 108k + 24 \times 84 > 0.$$

Solving these inequalities, $4 \le k < 24$ or $84 < k \le 144$. In these ranges, the k values of the form $20m + 14$ are: $14, 94, 114, 134$. Solve for the corresponding x values to get the final answers: $\sqrt{29}/2, \sqrt{189}/2, \sqrt{229}/2, \sqrt{269}/2$.

Problem 3.11 Solve for x: $x + \{x\} = 2 \lfloor x \rfloor$ $(x \ne 0)$.

Answer

1.5

Solution

Rewrite the equation to $\lfloor x \rfloor + \{x\} + \{x\} = 2\lfloor x \rfloor$, so $2\{x\} = \lfloor x \rfloor$. Since $0 \le \{x\} < 1$, and $\lfloor x \rfloor$ is an integer, we get $\{x\} = 1/2$, and $\lfloor x \rfloor = 1$. Therefore $x = 3/2$.

Problem 3.12 Solve for x: $x^2 - 4x + 2\lfloor x \rfloor^2 = 0$.

Answer

0

Solution

Use discriminant: treat the $\lfloor x \rfloor$ as a constant. For there to be real roots, $4^2 - 8\lfloor x \rfloor^2 \ge 0$, so $\lfloor x \rfloor^2 \le 2$, which means $\lfloor x \rfloor = 0, \pm 1$. Checking each of these values, the only one that works is $\lfloor x \rfloor = 0$, and then $x = 0$.

Problem 3.13 **(a)** Find all real numbers x such that $\{x\} + \left\{ \dfrac{1}{x} \right\} = 1$.

Answer

$x = \dfrac{1}{2}(k \pm \sqrt{k^2 - 4})$, $k = 3, 4, 5, \ldots$

Solution

Since $\lfloor x \rfloor + \{x\} = x$, and $\left\lfloor \dfrac{1}{x} \right\rfloor + \left\{ \dfrac{1}{x} \right\} = \dfrac{1}{x}$, we get $x + \dfrac{1}{x} = \lfloor x \rfloor + \left\lfloor \dfrac{1}{x} \right\rfloor + 1$. Thus $x + \dfrac{1}{x}$ must be an integer. Let this integer be k, then $x + \dfrac{1}{x} = k$, and $x^2 - kx + 1 = 0$. Quadratic formula yields $x = \dfrac{1}{2}(k \pm \sqrt{k^2 - 4})$. If $|k| = 2$, $|x| = 1$, and this does not fit in the equation. For $k \ge 3$, all solutions are in the form $x = \dfrac{1}{2}(k \pm \sqrt{k^2 - 4})$, $k = 3, 4, 5, \ldots$

(b) Show that any x that satisfies the equation above is irrational.

Solution

Since all solutions are in the form $x = \dfrac{1}{2}(k \pm \sqrt{k^2 - 4})$. When $|k| \ge 3$, the smallest difference between two squares is 5. Thus $k^2 - 4$ cannot be a square, so the solution is not a rational number.

4   Solutions to Chapter 4 Examples

Problem 4.1 Let $m > 1$ be an integer. What is the sum of all positive integers less than m and relative prime to m?

Answer

$$\frac{m\phi(m)}{2}$$

Solution

For $1 \leq a \leq m - 1$, if $\gcd(a, m) = 1$, then $\gcd(m - a, m) = 1$ as well, and $a \neq m - a$. Pair up all the a and $m - a$ that are relatively prime to m, and the sum of each pair is m. There are $\phi(m)$ such numbers, so the number of pairs is $\dfrac{\phi(m)}{2}$, therefore the total sum is $\dfrac{m\phi(m)}{2}$.

Problem 4.2 Let m be a positive integer. Find $\displaystyle\sum_{d \mid m} \phi(d)$. Here the sum is taken over all positive divisors d of m.

Answer

m

Solution

If d is a divisor of m, then the number of positive integers a between 1 and m such that $\gcd(a, m) = d$ is $\phi\left(\dfrac{m}{d}\right)$. If we categorize the numbers between $1 \leq a \leq m$ based on the values of $\gcd(a, m)$, namely put all a's with the same $\gcd(a, m)$ in the same category, then

$$\sum_{d \mid m} \phi\left(\frac{m}{d}\right) = m.$$

If d is a factor of m, then so is $\dfrac{m}{d}$, and vice versa. Thus

$$\sum_{d \mid m} \phi(d) = \sum_{d \mid m} \phi\left(\frac{m}{d}\right) = m.$$

Problem 4.3 Find all positive integers n such that $\phi(n) = \frac{1}{2}n$.

Answer

$n = 2^k$ for positive integer k

Solution

Let $p_1, p_2, \ldots$ be the prime factors of n. Since $\phi(n) = n\left(1 - \frac{1}{p_1}\right)\left(1 - \frac{1}{p_2}\right)\cdots$, we get

$$\left(1 - \frac{1}{p_1}\right)\left(1 - \frac{1}{p_2}\right)\cdots = \frac{1}{2}.$$

The only way for this to happen is that there is only one prime factor, and it is 2. Therefore $n = 2^k$, k is a positive integer.

Problem 4.4 Simplify: $\sum_{d|n} \frac{1}{d}$.

Answer

$\dfrac{\sigma(n)}{n}$

Solution

$$\sum_{d|n} \frac{1}{d} = \sum_{d|n} \frac{d}{n} = \frac{\sigma(n)}{n}.$$

Problem 4.5 Find the Möbius transform of $\mu(n)$.

Answer

$\delta(n)$

Solution

Let

$$g(n) = \sum_{d|n} \mu(n).$$

If $n = 1$, clearly $g(1) = 1$.

If $n = p^k$ where p is prime and k is a positive integer, then the factors of p^k are $1, p, p^2, \ldots, p^k$. By definition, $\mu(1) = 1$, $\mu(p) = -1$, and $\mu(p^k) = 0$ if $k \geq 2$. Therefore

$$\begin{aligned} g(p^k) &= \mu(1) + \mu(p) + \mu(p^2) + \cdots + \mu(p^k) \\ &= 1 + (-1) + 0 + 0 + \cdots + 0 \\ &= 0. \end{aligned}$$

Given that $\mu(n)$ is multiplicative, $g(n)$ is also multiplicative, and thus $g(n) = 1$ if $n = 1$ and $g(n) = 0$ if $n > 1$. This is exactly the delta function $\delta(n)$.

Problem 4.6 Find all positive integers n such that $3 \nmid \phi(n)$.

Answer

All n such that $9 \nmid n$ and n does not have prime factors of the form $3k + 1$ where k is a positive integer

Problem 4.7 Show that for any positive integers m, n,

$$\phi(mn)\phi(\gcd(m,n)) = \gcd(m,n)\phi(m)\phi(n).$$

Solution

Use the formula

$$\phi(n) = n \prod_{i=1}^{k} \left(1 - \frac{1}{p_i}\right).$$

The factors $\left(1 - \dfrac{1}{p_i}\right)$ appear once on both sides if p_i belongs to only m or only n, and appears twice on both sides if p_i is shared by m and n.

Problem 4.8 Find all positive integers n such that $\phi(n) = \dfrac{1}{3}n$.

Answer

$n = 2^k \cdot 3^l$ where k and l are positive integers

Solution

$n = 2^k \cdot 3^l$ where k, l are positive integers. Let $p_1, p_2, \ldots$ be the prime factors of n. Since

$$\phi(n) = n\left(1 - \frac{1}{p_1}\right)\left(1 - \frac{1}{p_2}\right)\cdots, \text{ we get}$$

$$\left(1 - \frac{1}{p_1}\right)\left(1 - \frac{1}{p_2}\right)\cdots = \frac{1}{3}.$$

Thus one of the prime factors must be 3, and other must be 2, and n has no other prime factors. Therefore the answer is

$$n = 2^k \cdot 3^l$$

where k, l are positive integers.

Problem 4.9 Let m and n be positive integers. Show that

$$\phi(mn) = \gcd(m, n)\phi(\text{lcm}(m, n)).$$

Solution

Clearly mn and $\text{lcm}(m, n)$ have the same prime factors. Let these prime factors be $p_1, p_2, p_3, \ldots$. Then

$$\phi(mn) = mn\left(1 - \frac{1}{p_1}\right)\left(1 - \frac{1}{p_2}\right)\cdots$$

and

$$\phi(\text{lcm}(mn)) = \text{lcm}(mn)\left(1 - \frac{1}{p_1}\right)\left(1 - \frac{1}{p_2}\right)\cdots$$

Comparing these two equations and using the fact that $mn = \gcd(m, n) \cdot \text{lcm}(m, n)$, we complete the proof.

Problem 4.10 Find the smallest positive integer k such that

(a) $\phi(n) = k$ has no solutions for n.

Answer

3

Solution

$\phi(n) = 3$ has no solutions.

(b) $\phi(n) = k$ has exactly 2 solutions for n.

Answer

1

Solution

$\phi(1) = \phi(2) = 1$. If $n > 2$, then $\phi(n)$ is even. Thus there are exactly 2 solutions $n = 1$ and $n = 2$.

(c) $\phi(n) = k$ has exactly 3 solutions for n.

Answer

2

Solution

$\phi(3) = \phi(4) = \phi(6) = 2$.

5   Solutions to Chapter 5 Examples

Problem 5.1 The equation
$$62 - 63 = 1$$
is obviously false. Can you move only one digit to make the resulting equation true?

Answer

$2^6 - 63 = 1$

Solution

Move the digit 6 is 62 to be the exponent of 2.

Problem 5.2 Can you find ...

(a) A multiple of 1350 that is a perfect cube? (You have *one second* to give an answer)

Answer

1350^3

Solution

The easiest way to get a perfect cube which is also a multiple of 1350 is clearly 1350^3. Multiple correct answers are possible.

(b) The smallest positive multiple of 1350 that is a perfect cube?

Answer

27000

Solution

$1350 = 2 \cdot 3^3 \cdot 5^2$, so the answer is $2^3 \cdot 3^3 \cdot 5^3 = 27000$.

(c) All positive multiples of 1350 that are perfect cubes?

Answer

$27000n^3$

Solution

Since 27000 is the smallest multiple of 1350 that is a perfect cube, all positive multiples of 1350 that are perfect cubes can be expressed as $27000n^2$ where n is any positive integer.

Problem 5.3 Diophantus was one of the last great Greek mathematicians; he developed his own algebraic notation and is sometimes called "the father of algebra." This riddle about Diophantus' age when he died was carved on his tomb:

> God vouchsafed that he should be a boy for the sixth part of his life; when a twelfth was added, his cheeks acquired a beard; He kindled for him the light of marriage after a seventh, and in the fifth year after his marriage He granted him a son. Alas! late-begotten and miserable child, when he had reached the measure of half his father's life, the chill grave took him. After consoling his grief by this science of numbers for four years, he reached the end of his life.

How long did Diophantus live? (Can you do it without algebra? Can you do it in three seconds?)

Answer

84 years

Solution

Let x be the age of Diophantus when he died. The riddle can be interpreted in algebra as follows:
$$\frac{x}{6} + \frac{x}{12} + \frac{x}{7} + 5 + \frac{x}{2} + 4 = x.$$

This equation becomes
$$\frac{25x}{28} + 9 = x,$$

which gives
$$x = 84.$$

To solve the riddle without algebra, we just need to notice that x is a multiple of both 12 and 7, and the only such integer that is meaningful as a person's age is 84, and we check that 84 in fact fit the riddle perfectly. Recognizing that the solution is 84 as the multiple of both 12 and 7 requires no more than 3 seconds for an experienced student.

Problem 5.4 Attach 3 digits after the number 503 so that the resulting 6-digit integer is a multiple of 7, 9 and 11. Find all such 6-digit integers.

Answer

503118 and 503811

Solution

The 6-digit number should be a multiple fo $7 \times 9 \times 11 = 693$. Since

$$503000 \equiv 575 \pmod{693},$$

we can attach $693 - 575 = 118$. Also $118 + 693 = 811$ is another choice. Therefore the two possible 6-digit numbers are 503118 and 503811.

Problem 5.5 What is the smallest positive integer ...

(a) ... that has exactly 10 factors?

Answer

48

Solution

Since $10 = 2 \times 5$, if a number has 10 factors, it must be of the form $p^4 q$ or p^9, where p and q are primes. Using $p = 2$ and $q = 3$, compare the numbers $2^4 \times 3 = 48$ and $2^9 = 512$, and clearly 48 is the smallest one.

(b) ... that has exactly 60 factors?

Answer

5040

Solution

Since $60 = 2 \times 2 \times 3 \times 5$, there are several possible cases for a number with 60 factors. Trying different possibilities, the smallest number with 60 factors is

$$2^4 \times 3^2 \times 5 \times 7 = 5040.$$

Problem 5.6 For positive integer $n > 2$, show that there must be a prime number between n and $n!$.

Solution

$n! - 1$ and $n!$ are relatively prime, so the prime factor of $n! - 1$ is what we want.

Problem 5.7 Let n be a positive integer greater than 11. Show that n must be the sum of two composite numbers.

Solution

Consider the three cases: $n = 3k$, $n = 3k + 1$, and $n = 3k + 2$, where $k \geq 4$.

If $n = 3k$, then $n = 6 + 3(k - 2)$. If $n = 3k + 1$, $n = 4 + 3(k - 1)$. If $n = 3k + 2$, $n = 8 + 3(k - 2)$.

Problem 5.8 Let $a_1, a_2, \ldots, a_{64}$ be a rearrangement of $1, 2, 3, \ldots, 64$, and let $b_1 = |a_1 - a_2|, b_2 = |a_3 - a_4|, \ldots, b_{32} = |a_{63} - a_{64}|$. Let $c_1, c_2, \ldots, c_{32}$ be a rearrangement of $b_1, b_2, \ldots, b_{32}$, and then calculate the 16 numbers $|c_1 - c_2|, |c_3 - c_4|, \ldots, |c_{31} - c_{32}|$, and so on, until there is only one number x left. Is x an even number or an odd number, or is it uncertain, depending on the process?

Answer

Even

Solution

For any two integers m and n,

$$|m - n| \equiv m + n \pmod 2,$$

therefore the sum of the resulting numbers at each step has the same parity as the sum of the previous step. Hence the last number remaining has the same parity as the sum of all the original numbers. In other words,

$$x \equiv 1 + 2 + \cdots + 64 \pmod 2.$$

Since

$$1 + 2 + \cdots + 64 = \frac{64 \times 65}{2}$$

is an even number, x must also be an even number.

Problem 5.9 Find the remainder when $10^{10} + 10^{10^2} + 10^{10^3} + \cdots + 10^{10^{10}}$ is divided by 7.

Answer

5

Solution

By Fermat's Little Theorem, $10^6 \equiv 1 \pmod 7$.

Consider the exponents of each term: 10^k for $k = 1, 2, \ldots, 10$. It is easy to verify that $10^k \equiv 4 \pmod 6$ for all $k = 1, 2, \ldots$, therefore each of the terms has the same remainder as 10^4 when divided by 7, which is 4. Thus

$$10^{10} + 10^{10^2} + 10^{10^3} + \cdots + 10^{10^{10}} \equiv 10 \times 4 \equiv 5 \pmod 7.$$

Hence the remainder is 5.

Problem 5.10 A perfect square can end with two 4s ($12^2 = 144$). Please answer the following questions.

(a) Can a perfect square end with three 4s? If yes, give an example; if no, prove it.

Answer

Yes

Solution

One example is $38^2 = 1444$.

(b) Can a perfect square end with four 4s? If yes, give an example; if no, prove it.

Answer

No

Solution

Assume

$$(100x + y)^2 \equiv 4444 \pmod{10^4},$$

so

$$y(200x + y) \equiv 4444 \pmod{10^4},$$

and y must be even. Let $y = 2z$, then

$$z(100x + z) \equiv 1111 \pmod{2500}.$$

This means z^2 ends in 11 ($\equiv 3 \pmod 4$), which is not possible for a perfect square.

Problem 5.11 Write down the integers from 1 to 1024 in reverse order: $102410231022\ldots\ldots 54321$. Perform the following "operation" on this number: take the first digit 1, multiply by 2 to get 2, add the next digit 0 to get 2, multiply by 2 to get 4, add the next digit 2 to get 6, etc., each time multiply by 2 and add the next digit, until the last digit 1 is added. Now the result is a large integer. Since this result has more than one digit, we perform the same operation as above on it, and keep going until the final result is a one-digit number. What is this one-digit number?

Answer

9

Solution

Suppose the number before the operation is a k-digit number, written in the place-value format (each of the a_i is a digit):

$$10^{k-1}a_{k-1} + 10^{k-2}a_{k-2} + \cdots + 10a_1 + a_0,$$

then the result we get after completing the operation is

$$2^{k-1}a_{k-1} + 2^{k-2}a_{k-2} + \cdots + 2a_1 + a_0.$$

Since $10^m - 2^m$ is a multiple of 8 for any positive integer m, the difference between this new number and the number before the operation is a multiple of 8. Therefore, the operation does not change the remainder when the number is divided by 8. Also it doesn't matter how many times the operation is performed, and the result always has the same remainder when divided by 8. Since the original number has remainder 1 when divided by 8, the resulting number should also have remainder 1 when divided by 8. At the end, the final one-digit number should have remainder 1 when divided by 8. Among one-digit numbers, only 1 and 9 have this property, and 1 is not possible to be the final result, so the answer is 9.

Problem 5.12 Find the last digit of

$$47^{47^{\cdot^{\cdot^{\cdot^{47}}}}},$$

where there are $k(>1)$ 47s.

Answer

3

Solution

$47^4 \equiv 1 \pmod{10}$. Also $47 \equiv -1 \pmod 4$, therefore the last digit is that of 47^3, which gives 3.

Problem 5.13 Find the remainder when 2^{345} is divided by 400.

Answer

32

Solution

We have

$$2^{10} = 1024 \equiv -1 \pmod{25},$$

so

$$2^{20} \equiv 1 \pmod{25}.$$

Since $345 = 20 \times 17 + 5$, we get

$$2^{345} \equiv 2^5 \equiv 7 \pmod{25}.$$

It is clear that 2^{345} is a multiple of 16. The only number under 400 that is a multiple of 16 and is also 7 $\pmod{25}$ is 32.

Problem 5.14 (2018 ZIML Master Round) Find the set of all positive integers n such that

$$1^n + 2^n + 3^n + 4^n + 5^n + 6^n$$

is a multiple of 7.

Answer

All positive integers except for multiples of 6

Solution

By Fermat's Little Theorem, $a^{6k} \equiv 1 \pmod 7$ for $a = 1,2,3,4,5,6$ and any positive integer k. Thus

$$1^{6k} + 2^{6k} + 3^{6k} + 4^{6k} + 5^{6k} + 6^{6k} \equiv 6 \pmod 7$$

for all positive integers k, therefore n should not be a multiple of 6.

If n is of the form $6k+1$, $6k+3$, or $6k+5$, we pair up the terms as follows: $1^n + 6^n$, $2^n + 5^n$, and $3^n + 4^n$. Clearly each pair has a factor 7, thus the total sum is a multiple of 7.

If n is of the form $6k+2$ or $6k+4$, we only need to verify that $1^n + 2^n + 3^n$ is a multiple of 7, since

$$1^n + 2^n + 3^n + 4^n + 5^n + 6^n \equiv 1^n + 2^n + 3^n + (-3)^n + (-2)^n + (-1)^n \equiv 2(1^n + 2^n + 3^n) \pmod 7.$$

Since

$$1^{6k+2} + 2^{6k+2} + 3^{6k+2} \equiv 1^2 + 2^2 + 3^2 \equiv 14 \equiv 0 \pmod 7,$$

and

$$1^{6k+4} + 2^{6k+4} + 3^{6k+4} \equiv 1^4 + 2^4 + 3^4 \equiv 98 \equiv 0 \pmod 7,$$

we get that $1^n + 2^n + 3^n + 4^n + 5^n + 6^n$ is divisible by 7 if $n = 6k+2$ or $n = 6k+4$.

In conclusion, $1^n + 2^n + 3^n + 4^n + 5^n + 6^n$ is a multiple of 7 if and only if n is not a multiple of 6.

Problem 5.15 Find the last three digits of

$$1 \times 3 \times 5 \times \cdots \times 1989.$$

Answer

375

Solution

By Chinese Remainder Theorem, the last three digits of a number is uniquely determined by its remainders when divided by 8 and 125. Clearly, the number $1 \times 3 \times \cdots \times 1989$ is a multiple of 125. Now we determine the remainder when divided by 8.

For any nonnegative integer k,

$$
\begin{aligned}
&(8k+1)(8k+3)(8k+5)(8k+7) \\
\equiv\ & 1 \times 3 \times 5 \times 7 \\
\equiv\ & 105 \\
\equiv\ & 1 \quad (\text{mod } 8).
\end{aligned}
$$

We also know that $1989 \equiv 5 \pmod 8$, thus

$$
\begin{aligned}
&1 \times 3 \times 5 \times \cdots \times 1989 \\
\equiv\ & (1 \times 3 \times 5 \times 7) \times (9 \times 11 \times 13 \times 15) \times \cdots \times (1985 \times 1987 \times 1989) \\
\equiv\ & 1 \times 3 \times 5 \\
\equiv\ & 7 \quad (\text{mod } 8).
\end{aligned}
$$

Since the product is an odd multiple of 125, the last three digits can be 125, 375, 625, or 875. Among them, only $375 \equiv 7 \pmod 8$. Therefore the answer is 375.

Problem 5.16 Old McDonald went to the Market and bought 100 chickens for exactly 100 dollars, among which roosters cost 5 dollars each, hens cost 3 dollars each, and three baby chicks cost 1 dollar. How many chickens of each type did he buy? Find all possible solutions.

Answer

$(0, 25, 75), (4, 18, 78), (8, 11, 81), (12, 4, 84)$

Solution

Let r be the number of roosters, h be the number of hens, and c be the number of baby chicks. Then

$$
\begin{aligned}
r+h+c &= 100, \\
5r+3h+\frac{c}{3} &= 100.
\end{aligned}
$$

This is a diophantine equation. Clear the denominator from the second equation,

$$15r + 9h + c = 300,$$

subtracting the first equation,

$$14r + 8h = 200,$$

simplifying,

$$7r + 4h = 100,$$

---

so
$$r = \frac{100 - 4h}{7}.$$

From this equation, r must be a multiple of 4, and $r \leq \frac{100}{7} = 14\frac{2}{7}$. Therefore, we get 4 choices for r: 0, 4, 8, and 12. The corresponding values for h are: 25, 18, 11, and 4. Then the corresponding values for c are: 75, 78, 81, and 84.

Therefore there are 4 groups of possible solutions:

$$(r, h, c) = (0, 25, 75), (4, 18, 78), (8, 11, 81), (12, 4, 84).$$

Problem 5.17 Show that 1599 is not the sum of 14 perfect 4th powers.

Solution

Consider the 4th powers of an integer n modulo 16. Let n be an even number, then $n = 2k$ where k is an integer, then

$$n^4 = (2k)^4 = 16k^4 \equiv 0 \pmod{16}.$$

if n is an odd number, let $n = 2k + 1$ where k is an integer, then

$$n^4 = (2k+1)^4 = 16k^4 + 4 \cdot 8k^3 + 6 \cdot 4k^2 + 4 \cdot 2k + 1 \equiv 8k(k+1) + 1 \equiv 1 \pmod{16}$$

because $k(k+1)$ is an even number.

Therefore, the only possible remainders for the 4th power of an integer are 0 and 1.

We know that $1599 \equiv 15 \pmod{16}$, so it is not possible to be the sum of 14 of values, each of which can only be 0 or 1 modulo 16. Thus 1599 is not the sum of 14 4th powers of integers.

Problem 5.18 Find the maximum positive integer n, such that $3^n \mid 2^{3^m} + 1$ for every positive integer m.

Answer

2

Solution

If $m = 1$, $2^{3^1} + 1 = 9$, so $3^n \mid 9$ and thus $n \leq 2$.

Now we show that for any positive integer m, $9 \mid 2^{3^m} + 1$. Indeed,

$$2^{3^m} + 1 = (2^3)^{3^{m-1}} + 1 \equiv (-1)^{3^{m-1}} + 1 \equiv 0 \pmod{9},$$

therefore $9 \mid 2^{3^m} + 1$. Hence the maximum possible n is 2.

Made in the USA
Middletown, DE
21 April 2022

64615079R00060